Real-World Software Projects for Computer Science and Engineering Students

Real-World Software Projects for Computer Science and Engineering Students

Varun Gupta and
Anh Nguyen-Duc

CRC Press
Taylor & Francis Group
Boca Raton London New York

CRC Press is an imprint of the
Taylor & Francis Group, an **informa** business

First edition published 2021
by CRC Press
6000 Broken Sound Parkway NW, Suite 300, Boca Raton, FL 33487-2742

and by CRC Press
2 Park Square, Milton Park, Abingdon, Oxon, OX14 4RN

© 2021 Varun Gupta and Anh Nguyen-Duc

CRC Press is an imprint of Taylor & Francis Group, LLC

ISBN: 978-0-367-63598-5 (hbk)
ISBN: 978-1-032-00253-8 (pbk)
ISBN: 978-1-003-11988-3 (ebk)

Typeset in Times
by codeMantra

Contents

Authors

Dr. Varun Gupta received his Ph.D. and Master of Technology (By Research) in Computer Science and Engineering from Uttarakhand Technical University and Bachelor of Technology (Hon's) from Himachal Pradesh University, respectively. He also holds an MBA (General) from Pondicherry University (A Central University).

He is working as postdoctoral researcher with Universidade da Beira Interior, Portugal. He is also visiting postdoctoral researcher, School of Business, FHNW University of Applied Sciences and Arts Northwestern Switzerland. He was honorary research fellow of the University of Salford, Manchester, United Kingdom (2018 to 2021).

He is associate editor of IEEE Access (Published by IEEE, SCIE Indexed with 4.098 impact factor), associate editor of *International Journal of Computer Aided Engineering and Technology* (Published by Inderscience Publishers, Scopus indexed), associate editor of IEEE Software Blog, associate editor of *Journal of Cases on Information Technology* (JCIT) (Published by IGI Global and Indexed by Emerging Sources Citation Index (ESCI) & SCOPUS) and former editorial team member of *British Journal of Educational Technology (BJET)* (Published by Wiley publishers, SCIE Indexed with 2.729 impact factor). He had been guest editor of many special issues published/ongoing with leading international journals and editor of many edited books to be published by IGI Global and Taylor & Francis (CRC Press).

He had organized many special sessions with Scopus Indexed International Conferences worldwide, proceedings of which were published by Springer, IEEE, and Elsevier. He is serving as reviewer of IEEE *Transactions on Emerging Topics in Computational Intelligence*. His area of interest is Evidence-Based Software Engineering, Evolutionary Software Engineering (focusing on Requirement Management), Business Model Innovation, and Innovation Management.

Dr. Anh Nguyen-Duc is an associate professor at the University of South Eastern Norway since 2017. He received his Ph.D. in Computer and Information Science at Norwegian University of Science and Technology from Trondheim, Norway. His current research interests include software startups, software ecosystems, AI ethics, and secure software development. He has more than 80 peer-reviewed publications in high-ranked journals and conferences. He has three edited books in business-driven software engineering. He serves as a chair of ten organization committees, a reviewer of recognized journals (including *Communications of ACM, Information Software and Technology (IST), Journal System and Software (JSS)*, and *Empirical Software Engineering (ESE)* and guest editor of several special issues in different software engineering journals. Recently, he has worked as a regional coordinator in an EU Interreg project valued at 4.2 million Eur. He is currently the chairman of the Software Startup Research Network (www.softwarestartups.org).

Introduction

1

This chapter gives the overview about what is present in the rest of the book. With the focus on achieving intended learning outcomes and constructive alignment, the chapter presents the importance of a project-based course in an engineering program (Section 1.1), describes basic building blocks of the course (Section 1.2), possible intended learning outcomes (Section 1.3), and practical consideration for educators and students when planning and participating in the course (Section 1.4). Moreover, the book introduces the concept of incremental project-based learning, which aims at running multiple instances of the course (Section 1.5). Last but not least, the chapter describes the target audiences and the overall structure of the book (Section 1.6).

1.1 MOTIVATION

Software engineering (SE) is one of the most demanding jobs in industry nowadays and its demand keeps on increasing. The software industry expects the engineering institutes and universities in general to produce students with industry-compatible competence and sound theoretical knowledge and highly practical skills. However, educating students to meet industry's demands has always been a challenge of ensuring practical competence while still providing future-proof research-based knowledge. In at least two decades from the report of Lethbridge (Lethbridge 1998), there exists a gap between university graduates' abilities and industry expectations, including both technical and nontechnical skills. Recent research has found that Computer Science and Engineering (CS&E) graduates still experience much difficulty transitioning into their new roles in industry (Valstar et al. 2020). Furthermore, students should be aware of recent trends in the software industry, such as continuous development and DevOps (Chen 2015, Fitzgerald and Stol 2014, 2017), Lean startup (Ries 2011, Nguyen-Duc et al. 2020), digital transformation, etc., which are likely to present their working contexts in the future.

In a typical bachelor program of CS&E, students undergo through theory subjects and laboratory exercises to gain practical understanding of the theory, and it is very hard to balance the theoretical foundation with abilities in current practices and technologies (Bruegge et al. 2015). Students need knowledge and awareness of CS&E principles, not only proper coding but also various SE knowledge areas, such as requirement, architecture, requirement, and testing. Besides, they should be able to understand and use tools, libraries, and frameworks – the integral elements of modern software development. When teaching aspiring software developers, one would ask a question: "what make these students future competent software engineers?" Possible answers might include the ease of coming up with sound technical solutions, but technical proficiency is no longer enough. The SWEBOS (Software Engineering Body of Skills) framework (Sedelmaier and Landes 2014) highlights the shortcomings of soft skill inclusions in the conventional SWEBOK (Software Engineering Body of Knowledge) model (Bourque et al. 2014). An academic program in CS&E might need to equip their students with ability to reason about and even develop new methods, competencies, and tools as needed (Jaakkola et al. 2006). However, there is no global consensus about what should be taught and how to teach to bridge the academia–industry gap.

With the lack of practical knowledge of implementing the challenging SE tasks, lack of understanding of the appropriate development methodology such as Agile, and lack of skills to employ software development tools, the experience of knowledge that can be glued together is missing as the prerequisite experience for undertaking the industrial live problems. For example, even if students know well the concepts of programming language such as C, concept of data structure, and concept of file handling and have implemented a small class projects using these concepts, but rarely knows the use of software development methodology, then the net output in the development of any complex project will be nondeterministic. This is because they might not be familiar with the practical realities of software development environment such as dynamism in user requirements, stakeholders, experience in handling stakeholders, etc. In case of lack of knowledge in implementing small projects, the practical knowledge of theoretical knowledge might be limited, and the situation will be very bad for such a student.

With every software industry working in a competitive market, efforts are to convince the stakeholders and clients by delivering the best-quality product that has a competitive edge. A student with the degree in any engineering branch is eligible to work in the software industry due to various roles required to be played by the company employees. The software industry requires candidates of different areas of interest, different expertise, and different experience, but with a high degree of self-motivation.

For example, a company producing a mass market product may require software engineers who can work as analysts, designers, database experts, programmers (such as mobile app developers and web developers), domain experts, and testers, and can be persons with good management skills to look at the overall management of development process, especially agile projects. Thus, every engineering student, whether interested in artificial intelligence, networking, SE, or programming, has a role to play in the software industry. A person with an interest in artificial intelligence may be required to work as domain expert and may contribute to the overall development of software for artificial intelligence projects. So, the interest of CS&E students or students of other branches does not come in the path of their selection in the software industry due to the demand of people of different expertise and capabilities.

When entering a CS&E course, one would expect that what they learn can be useful to their accumulated knowledge and skills can be utilized to make some practical outcomes. No matter if the course is about fundamental programming, such as Web Development, Object-Oriented Programming, Database, or advanced topics, such as Artificial Intelligence, Big Data, and Cyber Security, there are often individual or group assignments to produce concrete piece of code in the context of a mini-project, where project planning, execution, and closing are carried on. Students should be engaged in practical activities that they learn themselves the importance of both code and non-code activities, that will at varying complexities. A common approach in CS&E education context is project-based course, which is a small project designed by the lecturer and has students work in teams to complete the project by the end of the semester. Ideally students learn to work in teams and share responsibility for developing a codebase. The projects must be developed using appropriate methodology so that the student faces the development challenges very early and thus handling industrial projects becomes enriching experience for him. Students drive their own learning through inquiry, as well as work collaboratively to research and create projects that reflect their knowledge.

In a formal term, the "project-based learning" (PBL) represents a comprehensive educational approach that encourages students to provide solutions for nontrivial problems (Blumenfeld et al. 1991). The approach is aimed at the acquisition of fundamental knowledge, but it allows for the development of other desirable skills and attributes, such as using tools, group collaboration, and communication. The importance of PBL for engineering education is well perceived at the global scale (Bell 2010). Essentially, the PBL should set up a project setting that simulates as much as possible the industrial context. The guidelines reported in this book could serve for short-term as well as long-term academic objectives as per curriculum and individual career requirements i.e., iterations of PBL that last for the whole academic program or more.

Building a proper and practical project-based course is a nontrivial endeavor. It is a question to every educator that how can one adopt PBL principles to design, operate, and manage a course that is comparable to industrial live problems. Many challenges occur when building such course – it is difficult to find proper customers for students, it is difficult to adjust the requirements to balance the realistic and education of the project, it is difficult to establish a channel that students can learn from industrial customers, etc. The risk is that such project may be oversimplified, and students gain no experience interacting with clients or with code written by others. Establishing touch points with industries including expert visits, industrial internships, flexibility to undertake industrial assignments, participation in industrial challenges etc. seems to be a viable approach to build a project course. However, much needs to be considered for adopting PBL in the project course.

1.2 FUNDAMENTAL ELEMENTS OF PROJECT-BASED LEARNING

Depending on different educational program and intended learning objectives, students might be required to have prerequisite knowledge and skills prior to participating in a project-based course. In SE programs, course planners might expect that their students will experience several scenarios where SE knowledge areas, such as software requirement, software architecture, software construction, software testing, software project management, and SE process, can be practiced. The course by itself will typically be not designed to theoretically teach all SE knowledge. Therefore, SE concepts, techniques, processes, and tools are often prerequisites.

The most common element of PBL is teamwork: students work with their peers to solve a problem. The students should be able to understand the importance of good teamwork and team dynamics. Situations that the students are expected to face include handling difficult customers, coordinating team efforts, task distribution and responsibilities, and collective problem-solving. The students will thus obtain skills related to team communication, task management, collective decision-making, team retrospection, and leadership. In practical project settings, we focus on practical problems, such as real customers or simulated situations of real-world problems. Customers and projects are the key setting for this type of courses. As an educational course, pedagogical elements, such as lectures, supervision, and grading, are also discussed as a part of the course setting (Figure 1.1).

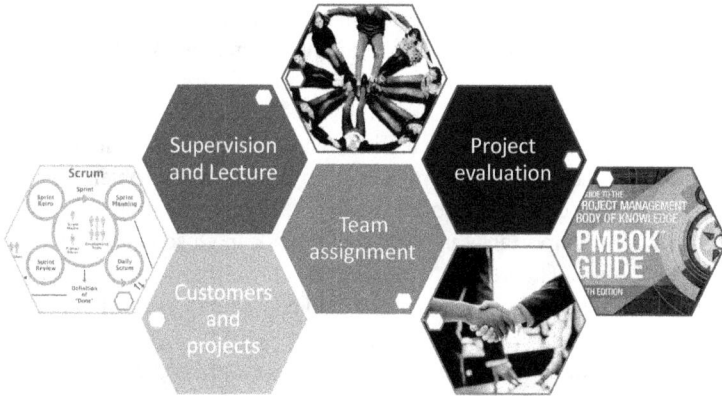

FIGURE 1.1   Course setting to facilitate the acquisition of learning outcomes.

1.3 LEARNING OBJECTIVES

The two key features of project-based learning are: (1) the presence of a problem or a question, which serves to drive activities that result in a series of artifacts or products that (2) culminate in a final product which addresses the driving question (Helle et al. 2006). From a problem as the starting point, the courses should be designed to direct student's activity toward effective learning, and the project implementation process is at least as important as the project outcome is. A learning objective, also called a learning outcome, describes what students should be able to demonstrate the result and how the student should show the attainment of these goals (Biggs and Tang 2011). From the authors' experience, we describe in our courses intended learning outcomes based on three areas of learning: knowledge, skills, and general competence. In general, for a project course, the learning outcome should be:

- Knowledge: to understand different engineering and management aspects of a software development project and to reflect on success and failure factors of the project from individual and team levels
- Skills: to organize and implement concrete prototypes of CS&SE products or services through executing all phases of team-based software development projects, as well as document and present results to a realistic client

- General competence: provide insight into project work and how groups can be used to solve complex computer science and SE problems

Each group is given a task from a client that is to be carried out as a project. All phases of a development project are to be covered: problem discovery, requirements specification, solution design, implementation, testing, and evaluation. It is important that the groups work in close collaboration with actual customers. The groups will hand in a project report and give a final presentation and demonstration of a runnable system to the customer and the censor.

In a particular setting, i.e., a course that teaches Agile development or a course that teaches software security, the learning objectives can be made more detailed. Three major areas of learnings are expected after completing the course: (1) specialized knowledge, (2) soft skill, and (3) teamwork. A project course might include some or all detailed objectives as below:

Regarding *Project Management*, the team should be able to:

- Perform fundamental project planning activities, such as project planning, management of scope, time, cost, effort, and risk.
- Conduct project documenting with regard to all delays, overruns, and weaknesses, so that they can be explained and augmented.
- Execute their own project plans with monitoring of their activities, efforts, and resource usage.
- Finalize the product implementation, preparation of final deliveries, i.e., prototype, manual, presentations, etc.
- Deliver and present the final prototype to the customer/external examiner. Under the final presentation and demonstration, it is important to give the customer a complete and good impression of the system delivered.

Regarding *Teamwork*, after the course, the students should be able to:

- Assign, coordinate efforts, and distribute work and responsibilities among team members.
- Understand the importance as well as challenges of having a good team collaboration.
- Experience collective decision-making, task assignment, and conflict resolution. The students would experience collaborative working environment. Earlier in your studies, the assignments have been smaller and more well-defined. In this project, there are (conflicting) decisions to be made.

- Perform various soft skills including problem-solving, creativity, decision-making under uncertainty, and leadership.
- Realize and adapt to no-ideal working situations.

Regarding *SE*, after the course, students should be able to:

- Handle normal and difficult customers. They can be unreliable and/or unavailable. They might change directions, come up with new ideas, and have an unclear picture of what they really want. An important part of this course is to manage the group project, so that the results match the customer's needs, even though the situation may turn difficult.
- Understand the process of requirement explication.
- Document requirements properly, i.e., as user stories, putting them into Sprint Backlogs
- Provide architectural solutions for the given task, visualizing them as architectural diagrams, i.e., use case diagrams, architectural views, database diagrams, workflow diagrams, etc.
- Apply programming skills with both front-end and back-end development into a real project.
- Provide a test plan to ensure the quality of delivered products. The test plan might include test scenarios, test cases, and test reports at different levels of testing.
- Document the whole project report. The final project documents must be complete, well-structured, and target the technical knowledge level of the customer.

Regarding some *specialized topics*, for instance, Artificial Intelligence projects, Global Software Development, or Lean Startup courses, students are expected to be able to:

- Demonstrate that they are able to plan and manage small SE projects using agile methods such as Scrum and XP.
- Apply their SE knowledge to participate in a software project in a distributed setting.
- Explain, identify, and apply security mechanisms implemented in iOS and Android mobile application platforms.
- Gain hands-on experience designing and implementing relatively large AI projects.
- Gain valuable insights into why, when, and how to use AI methods in realistic problems that they may encounter in their technical careers.

1.4 CHALLENGES WHEN DESIGNING A PBL COURSE

Several shortcomings with PBL approach within CS&E education are acknowledged in education literature. First, students are often unprepared for the scope of a practical project given the more limited nature of their preparation for such experience (Albanese and Mitchell 1993, Hyman 2001, Helle et al. 2006). Second, most project course is adopted in a cross-sectional manner and lacks longitudinal and progressive setting. This leads to the fact that in most of the project courses, students do not have chance to refine their products, artifacts, and practices, which is an important component within the profession. Third, having an appropriate set of requirements as inputs for students is not an easy task. Ambiguous requirements or tasks often lead to different interpretations among students, which lead to mistakes and confusion and consequently wasted efforts. For non-experienced customers who are not familiar with the course (a researcher or a non-IT officer), it is often difficult to understand what customers really want in the beginning. Fourth, customer inputs come at different levels of complexities and difficulties. Projects need to be set up considering the current background and knowledge of students. Fifth, customers might not have sufficient time to spend on the project, giving the limitation on how much industrial knowledge, skills, and practices can be present to students via the project setting. We experienced all these challenges in 10 years of teaching PBL courses. In later sections, we will introduce our best practices and recommendations for planning and designing such course.

1.5 INCREMENTAL PROJECT-BASED LEARNING

The student should be highly flexible to undertake real software development under extreme levels of market uncertainties and achieving expertise in different software engineer roles ranging from domain experts, to requirement analysts, designers, testers, and Research & Development (R&D) experts. The experience in handling the software development activities right from the beginning of the engineering education helps to incrementally transform the SE skills of the student and equip him to gain lessons from the rapidly changing nature of the software artifacts due to market dynamisms. The PBL is not one-time activity imparted in one course, as one capstone project, or as outcome of individual courses. A project-based course should offer a setting that students feel interested and motivated to produce the intended outcomes. The motivation could be intrinsic or extrinsic. If students are intrinsically motivated, then providing them good experience with the real projects, making them able to transform their ideas into working software, simulating real software development environment in academic settings (for instance, allowing students to have real interviews with the real customers), etc., could be meaningful tools to foster motivation. Extrinsically motivated students require the project team to set aside the marks, grades, and other extrinsic motivation factors to motivate the students to work in the teams to complete the projects. The good mix of intrinsic and extrinsic motivation factors could be a booster for fostering student involvement in PBL.

Further, the students these days have interests that are interdisciplinary in nature. For instance, they want to solve interdisciplinary problems (for instance social problems) using computational solutions. Implementing this project requires strong collaboration among multiple specializations and may be difficult to be implemented in "one go" as the students have to gain expertise in multiple interdisciplinary fields. In these circumstances, the students could start the project early and could implement it in an incremental way, with each increment being continuously innovated with the new learning students got with the previous project implementation.

In academic portfolio, education programs have more than one project-based course which basically overlaps with each other. For instance, bachelor's degree in computer science may offer programming with data structures in C and programming in Java. These two courses overlap except that they are based on two different languages. The students will find it an ambiguous experience working on these project-based courses. To avoid this, these two projects could add to the incremental learning experience of the students. To accomplish this, students first implement their projects using C language in a particular semester. In the following semester, they try to innovate their project using Java language to take advantage of the features specific to this language and enhance the value to the customers. This avoids the repeated learning by continuously fostering the enhanced learning experience of the students. From a program perspective, coordination among these courses to make them enhanced learning iterations will make new learning for students.

In an incremental project, the students continuously search for the innovative ideas that could enhance the value to the customers (even though the customer is internal to the academic institution). The product is continuously innovated, and such innovation (generating ideas and their implementation) provides real experiences to the students. The innovation may not have positive impact in the market but as such projects are implemented in the academic settings, they bring a lot of learning to the students.

1.6 TARGET AUDIENCE AND STRUCTURE

This book is about reasoning, designing, and running a practical software project-based course from multiple perspectives. The book combines state-of-the-art knowledge about pedagogical approaches and empirical evidence running such course around the world. The book also contains the author's experience of more than 10 years of teaching customer-driven project courses at several universities. The book aims at contributing to address the industry–academia gap in one of the most common course settings. Typical audiences of the book include, but not limited to:

- Educators running CS&E project-based courses with a focus on sustainable and highly transferable competences.
- Educators running capstone course aiming at improving their connection to industry.
- Educators who strive to provide students a learning experience that paves the way for their smooth onboarding in industry.
- Educators who look for building blocks and design patterns for a project-based course and its curricula.
- Course and curriculum designers in search of approaches for integrating supervisors, lecturers, and customers' activities in student learning experience.
- Innovators who build a PBL experience into a year-long incremental academic offer.
- Students who look for ABC guideline into a CS&E project-based course.
- Students who are working in a project-based setting and want to reflect project activities.
- Students who organize and execute a project by themselves beyond the scope of academic program.

This book is structured into three sections. Chapter 1 contains the introduction to the book, from where you see why you should read this book and what you can learn from it. Chapter 2 presents common building blocks for a project-based course and different considerations when designing the course. Chapter 3 starts presenting insight of a project by giving definition, guideline, and discussion about project planning. Chapter 4 continues by briefly going through different execution processes and detailed examples on implementing a customer-driven project. Chapter 5 tells about matters when closing the student project, such as project delivery, final presentation, and reflection. These chapters help the student to accurately plan the project and execute the plan using SE model (for instance, agile scrum model) and finally close the project by reporting it. Within Chapter 5, we also provide a detailed structure of a recommended project report.

The book also recommends that the PBL in the academics could be formed as an incremental process, which allows the student to start from the scratch under high level of uncertainties and incrementally build improved version of the software solutions as a result of implementation of the validated learnings brought because of exposure to the real facts, feedbacks, and better insights brought with the passage of time. Chapter 6 presents the extension of the project-based course setting according to the incremental learning approach. It gives the ideas how project planning, execution, and closing activities can be adjusted to fit in a 3-year-long academic experience. Chapter 7 mentions the guidelines for dissemination of the project-related artifacts such as research, project code, experience, opinion, empirical evidences, etc., with the audience by selecting suitable publication venue using suitable reporting format. In the end, the appendix gives brief description of the various implemented projects and sample formats of various project-related artifacts that students could use in academic projects.

We note that some sections might be specially relevant and interesting for different targeting audiences. As described in Figure 1.2, educators and innovators might focus on planning and development of the course's curriculum while students would be interested in how to use the book as a guideline to implement their projects. Moreover, research-driven innovators can exploit the setting in a longer perspective and plan for scientific dissemination.

Targeting audiences   **Particular interest in ...**   **where to find ?**

Targeting audiences	Particular interest in ...	where to find ?
		Chapter 1
Educator	Why should I read this book?	Section 1.1
	What to plan when building the course?	Section 1.2 - 1.4
	Who such course is made for 3 years?	Section 1.5
		Chapter 2
	I want more detail about building the curriculum for the course	Section 2.1 - 2.3
	How can I judge project performance?	Section 2.4
	What I need to concern when running the course?	Section 2.5
Students		Chapter 3
	What I need to do when planning a project?	Section 3.1 - 3.5
	What kind of practices available for my team?	Section 3.6
		Section 3.7
	Ideas about tools and infrastructures for my team to start with?	Chapter 4
	Met the customer! What to start first ?	Section 4.1
	How teamwork works?	Section 4.2 - 4.3
		Section 4.4
Research-driven Innovator	What to do with Software Engineering in our project?	Chapter 5
	How to deliver properly?	Section 5.1
	Is there some project template and guideline that we can use directly?	Section 5.2
	I want to innovate the course to a three-year learning experience	Chapter 6
	What are other possible outcome?	Chapter 7

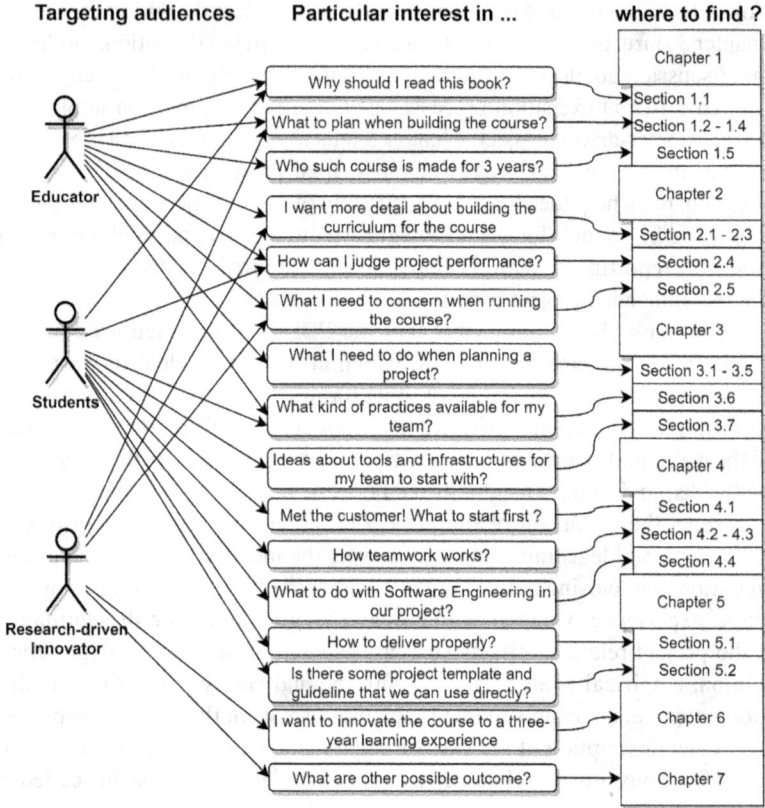

FIGURE 1.2   Mapping of audiences' interests and book structure.

Design dimensions of project courses

2

Project-based learning needs a setup that balances between flexibility and guidance in each dimension of the course. With too much freedom, the students' learning processes could quickly degenerate when encountering decisions and choices beyond their competence. However, a too strict process and hard constraints hinder innovativeness, stifle the competitive spirit among the groups, and may inadvertently change students' behaviors from controlled learning by doing, and failing, to rule following. This chapter presents different design dimensions of a project-based software engineering (SE) course and shares our best practices, including customer and projects (Section 2.1), teamwork (Section 2.2), supervision and lectures (Section 2.3), and course evaluation (Section 2.4). The detailed discussion about challenges when designing these dimensions is given in Section 2.5.

2.1 CUSTOMERS AND PROJECTS

Each project will have a group of representative customers, in most of the case the customer is the person who knows the best about the project. This section presents a background of customers and projects we have worked with over last 10 years.

2.1.1 Customers

Customers are one of the stakeholders of the project that have the intention to buy the software product. Customers could be same as the software users or different. In other words, customers may buy the product for themselves as end users (for instance, people buying the games for the use), or they may buy the software to be used by other people (end users) (for instance, banks buying mobile applications to be used by their clients). In remainder of this book, it will be assumed that customers are the actual end users of the software product, i.e., customer and user terms will be used interchangeably.

In practice, customers have the problems for which they require solutions. Thus, the software product will be purchased by the customers if it actually solves their problems and provides them benefits (i.e., higher perceived customer value). The product must match with the expectations of the customer, i.e., it must have product/market fit; otherwise, there will be no market for the product. In our course, industrial customers participate in academic courses to gain some benefits. Typical expectations include gaining access to students for the purposes of recruiting and getting direct or indirect benefits from the project outcomes.

Getting a real customer, especially an external one from industry (or from real social life), is not trivial and requires a substantial amount of work. First, we need to find the interested customers, agree with the customers on suitable project topics, then define a collaboration process between students and the customers to accomplish the work. Finally, we need to make sure that both the learning goals of the students and the goals of the companies get fulfilled. For an established course with a long operational history, acquiring customers might be less problematic. There are various types of customers, from research institutes, SMEs, and software departments at large companies who possess a problem that needs a student project to validate. The problems could also identify from the experiences of the teaching faculty, real-life problems, and experiences of the students for which they believe that solution is really required. The real-life problems could be solved by providing technologically advanced solutions by following SE principles in academic environment. Such social–technological solutions provide benefits to the software team in terms of their better understanding with the problem domain, better access to the customers facing the problems in real environment, and motivations for contributing toward society.

Customers can also occur from some special scenarios, as described in Table 2.1. Ideally, customers should attend the course as real customers instead of merely issuing a problem statement. One or two customer representatives will involve in the student project actively. The candidate customers

TABLE 2.1   Real customers and where to get them

TYPICAL CUSTOMERS	HOW TO APPROACH
Existing customers from of the courses or relevant courses	Documents from the previous instance of the course
Administrative, scientific, and educational staff in the same workplace	Teacher's scientific and educational workspace
Companies, public units in the teachers' professional network	Teacher's professional network or relevant lecturer's networks
Industrial customers in the regional	Reach out via regional events, job fair, emails to company contact lists
Alumni students of the course or the educational program	Reach out via student alumni contact list
	Recommended by colleagues
Entrepreneurs with an idea needed validation	Coworking spaces, incubator, accelerator programs
	Students from other programs in the school
Friends/family/acquaintances to the students	The students in the course can explore their close relationships with customers
Global customers from sibling course	Students and teachers from a sibling course aboard can be customers (exchanged customers)

need to submit their project proposals to be selected for the course. When there are more customers than the number of student teams, project selection needs to be performed.

The customers' involvement in the software development is very much required as their participation enhances the success rates of the project as they help to provide their inputs for problem domain and validation activities (to ensure that product matches their expectations). Here are honest reflections from our previous students about their customer participation:

> Throughout the project, the customer has been easy to work with. They listened to the team's requests and concerns, as well as gave a good balance between freedom and restrictions. Sometimes, if needed, they were open to adjust their requirements accordingly. This released some of the pressure on the team.

(Group2_ 2017)

Inge was very helpful to us throughout the project, and he spent a lot of time and effort in helping us. He arranged a private guide for us within the first week of the project and gave us free access to the cathedral whenever we wanted it. KUNDE provided a room we could use at our will. All the contents of the app were also provided by NAVN, always when we needed it. We arranged a meeting with the customer almost every week, even more often in some weeks towards the end of the project

(Group6_ 2015)

The basic structure for the collaboration between a customer and a student team is often defined by the software development process used. The detailed discussion of this process is given in later parts of the book, i.e., Sections 3.6 and 4.3. Customer expectation should be specified in the way to provide guideline for evaluating the project performance in the end. The initial project description can vary from vague to detailed content, for examples, three project descriptions in Appendix 1.

2.1.2 Projects

The students typically engage in all three phases in their projects, which are planning, execution, and closing. In the planning phase, the students get to know their team members, customers, and their requirements. Typically, the students voluntarily decide their roles and areas of responsibility. Most of the teams adopt a flat structure. They also make a preliminary project plan and set up the working environment. In the execution phase, projects are often divided into Sprints with frequent deliveries to customers. Student projects have to demonstrate SE activities (i.e., requirement elicitation, system design, coding, and testing). In the closing phase, the project results are submitted, demonstrated, and presented.

Based on our experience with student projects, defining a proper solution and problem domains are critical for students to be able to successfully complete the project. Problem domain refers to all information that defines the problem and constrains the solution (the constraints being part of the problem). It includes the goals that the customer wishes to achieve, the context within which the problem exists, and all rules that define essential functions or other aspects of any solution product. It represents the environment in which a solution will have to operate, as well as the problem itself.

In terms of SE, understanding the problem domain will typically start with the given project goals and descriptions, followed by the process of requirement elicitation, specification, and elicitation. When collecting user stories or user requirements, students should be able to decide which requirements are relevant for the projects. Not all of the wish list from the customer should transfer directly into a user story. It is recommended to keep in mind some higher-level statement of what the objectives are and what people think the problems are (if people didn't think there were problems, they wouldn't be paying someone to come up with a solution, would they?). Then it's either in, out, or borderline. If it's borderline, one probably wants someone to agree that it's out (even if you want it to be in). In the report, the students should be able to describe what is in and what is not in the scope of the project. Usually, a list of prioritized product backlog would tell the overall scope.

Stated in other words, depending on the source of the problems (whether it's industry, faculty, or real-life problems), the students need to explore the problem domain to enhance their understanding with the problems by interacting with the customers, observing the events occurring in the real context, etc. The literature is a meaningful source of providing the students with the research problems, especially through the support of empirical research studies especially case studies, surveys, systematic mapping studies, and systematic literature reviews (Gupta et al. 2015, 2020a–c). The students could use the empirical studies as the meaningful pointers for working on various problems in close collaborations with the customers that face such problems.

The continuous involvement of the customers for exploring the problem domain enhances their understanding of the problem domain, and continuous interactions with the customers help them further narrow down the problem domain. This also helps them to accurately (even if not with 100% accuracy) identify the solutions for the problems identified by the team. There could be multiple technical solutions for the problem in hand, but the technological familiarity of the team plays an important role in selection of suitable solution. This task could be termed as mapping of the problem domain (pains and gains) into a set of software features (along with their implementation details) that will be the part of the software under development.

While the problem domain defines the environment where the solution will come to work, the solution domain defines the abstract environment where the solution is developed. The differences between those two domains are the cause for possible errors when the solution is planted into the problem domain. In respect to a given problem (or set of problems), the solution domain (or solution space) covers all aspects of the solution product, including:

• The technical aspect of the solution product itself
• The process by which the solution is arrived at
• The environment in which it is constructed

In a project context, the solution domains are heavily impacted by the technical, process, and organizational capacity of the students to address the given problem. It could be that the students do not know about the required programming language to implement the products. It can also be that the students are not aware of the available cloud services that need to be integrated into the project system.

From our experience, all of our student teams spend significant time to explore both their problem and solution domains. The exploration is seen as an informal R&D process, involving a lot of self-learning. The team often reports this pre-study stage before describing their implementation of the solution.

It is worth noticing that every project is a unique experience. Comparing them is a difficult, not even infeasible task. Nevertheless, to provide a basic guideline for assessing the performance of students, the projects need to be put in a common framework. This is not fair if students working with simple projects will be graded in the same way with the students working with complex projects, in a common academic context. We utilized a simple sensemaking framework to describe a way to categorize student projects into different domains, as seen in Figure 2.1. Uncertainty represents the situation of imperfect or unknown information. Uncertain problem means that the customer does not have a clear vision on what they want to achieve. Certain problem is where customer can specify their needs in an implementable, precise specification. Uncertain solution means that the technologies, algorithms, architectures, processes, etc., to implement the products are (partly) unknown to the students. Certain solutions can happen when the students

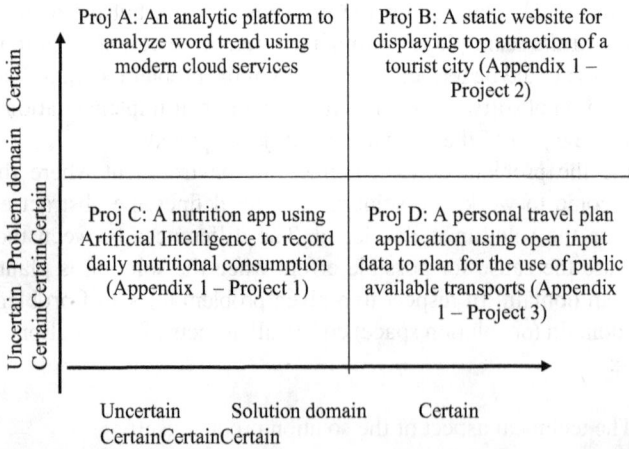

	Proj A: An analytic platform to analyze word trend using modern cloud services	Proj B: A static website for displaying top attraction of a tourist city (Appendix 1 – Project 2)
	Proj C: A nutrition app using Artificial Intelligence to record daily nutritional consumption (Appendix 1 – Project 1)	Proj D: A personal travel plan application using open input data to plan for the use of public available transports (Appendix 1 – Project 3)

Problem domain — Uncertain ... Certain

Solution domain — Uncertain ... Certain

FIGURE 2.1   Project types.

have done similar task before and they have hands-on experience to carry the work again. Figure 2.1 presents four examples of the four types of student projects (detailed project descriptions are provided in Appendix 1). We suggest that lecturers should consider an assessment of project proposals and eventually rank them according to their required technologies and complexities and requirement and solution uncertainties. The classification of the projects can be ad-hoc or simply noted by the teachers when assigning the projects to students. However, they can also be thoroughly conducted with the analysis of project's dimensions. The characteristics of projects should also be considered when evaluating their performance.

Tip 1: Projects should be classified basing on their uncertainty and complexity in both problem and solution domains. The evaluation of the student performance should consider their project types.

2.2 TEAMWORK

Teachers have several options to establish the setting for student teamwork. These includes (1) team size, (2) team assignment, and (3) team roles.

2.2.1 Team size

Teams, like any other organizational unit, need adequate staffing in terms of both quality and quantity of personnel. A team's work performance depends on its ability to efficiently and effectively work in a directly interactive mode to achieve a common team output. Besides many qualitative attributes of teams, such as leadership, management, communication, team size matter. In professional world, the concern of team size comes up any time a new team needs to be formed or an existing team is being evaluated. A professional team size can vary from two members to 20+ people per team. According to Hoegl, team size can affect teamwork in several ways (Hoegl 2005):

- Sharing of technical and coordinative information within the team becomes significantly more difficult as the number of team members increases.
- Larger team size also creates a stronger need to coordinate the contributions from the various team members.

- Team size is an important determinant of the social loafing phenomenon, whereby individuals decrease their effort as the number of people in the group increases.
- As the size of a team increases, so does the number of "nonparticipating" members.

The right team size depends on the work to be performed with some tasks requiring more team members than others do. Team size must be determined with respect to both staffing requirements, deriving from the size of the project task, and teamwork requirements, deriving from task complexity and uncertainty. A large team size might led to a less cohesive group, with less efficient communication and less information exchange among members. Large teams may also facilitate isolation and inactivity of some students. From our experience, a typical team size in project-driven course might range from five to eight members per team. From 2015 to 2020, we assigned students to groups from six to eight members.

A famous law in project management says that adding man power to a late software project makes the project later (Brooks 1995). Brooks defines a concept "ramp up" time – the time needed for an added person to become productive. Software projects are complex engineering endeavors, and new workers on the project must first become educated about the work that has preceded them; this education requires diverting resources already working on the project, temporarily diminishing their productivity while the new workers are not yet contributing meaningfully. Each new team member also needs to integrate with a team composed of several engineers who must educate him or her in the current repository gradually. In addition to reducing the contribution of experienced workers (because of the need to train), new team member may even make negative contributions, for example, if they introduce bugs that move the project further from completion. Recent research still reveals that larger teams are associated with decreased performance of individual students, poorer and less diverse social interactions, and this can also happen in an online context (Saqr et al. 2019).

Tip 2: Team size should be kept between five and eight members, keeping in mind that some students might drop out of the groups.

2.2.2 Team assignment

Table 2.2 represents three approaches to assigning students to teams that have been explored in educational science (Bacon et al. 2016): self-selection,

TABLE 2.2 Three common team assignment approaches

Random assignment	Members generally do not know each other beforehand
Teacher-assigned team	Instructor uses his or her judgment to assign team members to a particular project
Self-selection team	Students choose their team members by themselves

random assignment, and teacher assignment. Self-selection has been recommended by many educators because it may offer higher initial cohesion. Student teams often have very short longevity, perhaps only a few weeks to few months, and so the initial cohesiveness that self-selected teams often possess may help these teams to become productive more quickly. Self-selection may also encourage students to take more ownership of group problems motivating students to manage interpersonal conflict more successfully. Self-selection is not without problems, however, including the tendency for self-selected teams to be overly homogeneous and thus not offer the advantages that some diversity may provide. Self-selected teams may also possess an inadequate skill set, unless measures are taken to constrain self-selection. Thus, self-selection trades a possible lack of diversity and critical skills for initial cohesiveness and established norms. This concern can be addressed with teacher-assigned teams.

Teacher-selected teams are formed when the instructor uses his or her judgment to assign team members to a particular project. Teams may be formed based on interests expressed by the students in particular projects, geographical location, or student age and work history. The instructor can ensure a diverse group of students with regard to, for example, academic performance, location, gender, and nationality. Teacher-selected approach can be used in combination with self-selected approach to combine the cohesion and diversity into the team.

In some context, random assignment might be the option. In practice, it is not always the case that one will be able to select his/her own team. It is typical that one participating in an existing team or joining in a group of new colleagues. Each student begins the class with the same chance of working with every other student, but due to the random nature of this approach, the final team assignments can be quite unbalanced in terms of skills, diversity, and general ability. Random assignment is also not likely to generate teams with a useful combination of skills or create groups of students who want to work together.

Tip 3: Several team assignment strategies can be combined to facilitate a team setting for first-year students

2.3 SUPERVISION AND LECTURES

The course staff consists of the course teacher and several supervisors, who are previous students of the course. Each supervisor typically guides teams in issues related to the course regulation, development processes, and teamwork. A supervision meeting is held regularly. Its purpose is to keep the supervisors up to date on the team's work, to ensure the project does not fall behind on where it should be at given point. The team could also get feedback on the report and the deliveries expected for the course and suggestion for improvements if needed.

Before the projects begin, there are a few lectures related to the course arrangements, topic presentations, and used software development process. An example of how course lectures are organized is presented in Appendix 3. During the project the teacher can arrange sessions where students can present their experience and discuss with other teams.

2.4 COURSE EVALUATION

Since each student spends about 20 hours per week on the project in one semester, a typical project involving six students corresponds to about 2000 person-hours. The project is important for students' transcripts, as it carries 15 ECTS. The project work is evaluated on the basis of the quality of the project report, the functioning system prototype, the presentation delivered at the end of the course, and team dynamics. Customers consider their experience working with the team, the value of their output, and the delivered report as inputs for their evaluation. Supervisors observe their teams throughout the course to evaluate the teams' performance and learning. The customer and the supervisor are involved in an evaluation meeting that gives the final mark for the students.

Assessment rubric is a documented approach for systematic evaluation of student work. It tells what elements of student performance matter most and how the work is to be judged. The rubrics are often written to guarantee proper understanding of the expectations among the examiners resulting in fair assessment. Rubric-based assessment is successfully adopted in many SE courses (Petkov and Petkova 2006, Feldt et al. 2009). We should note that rubrics are not checklists. They are associated with the development of criteria and rating scales for evaluation of the product against these criteria. With the rubrics and the associated criteria, an examiner should be able to carry on a systematic evaluation of the student work and, ideally, provide a repeatable evaluation.

A possible scheme for an assessment rubric may include, but not limited to: Planning, Execution, Originality of the design, Use of resources, Critical review or self-assessment, Personal contribution, Comprehension of concepts and aims, Background information, Initiatives, Motivation/application, Appropriateness of methods and/or experimental design, Organizational skills, Competence and Independence, Ability to problem-solving (Heywood 2000). The degree to which the content of the project is taken into account or whether the assessment focuses on technical, process, or interpersonal issues depends on the course and the teachers. As shown in Figure 2.2, assessment rubric can base on concrete outcome of the projects and inputs from both supervisors and customers. In our recent course, we use three simplified dimensions:

1. The quality of the final project report: how complete it is, how correctly and honestly it reflects the requirement, design, implementation, and testing activities.
2. The quality of teamwork, process, and practice reports: how the team works together, to what extents Agile practices and tools are adopted.
3. The quality of the final Minimum Viable Product (Nguyen-Duc and Abrahamsson 2016, Nguyen-Duc et al. 2020): the complexity of the project, how much has been implemented against customer requirements? The result of user testing.

The detail of the rubric is developed as shown in Table 2.3. The rubric is successfully used in three courses' instance, although different instances have slightly adjustment from their ways to aggregate evaluations across these criteria. In our most recent course, the weights are given equally among the above criteria.

FIGURE 2.2   Four artifacts and two perspectives in a possible course rubric.

TABLE 2.3 Assessment rubric and detail criteria using in the course USN PRO1000

CRITERION	DEFINITIONS OF RUBRICS' SCALES		
	BEGINNING	DEVELOPING	ACCOMPLISHED	EXEMPLARY

1. The quality of the final report

1.1. Is the project planning sufficiently documented?

1.2. Are the problem space and solution space sufficiently studied and reported?

1.3. Are the development methods and their rationale sufficiently described?

1.4. What is the quality of the product backlog?

1.5. How is the product design written?

1.6. How is the testing written?

1.7. How is the Sprint summary written?

1.8. Is there sufficient reflection on both student and teacher's sides?

1.9. How is the report formatted and layouted?

1.10. Are the report and relevant material properly submitted?

2. The quality of teamwork and team process

2.1. How is teamwork in the beginning of the project?

2.2. How is teamwork in the middle of the project?

2.3. How is teamwork in the end of the project?

(Continued)

TABLE 2.3 (Continued) Assessment rubric and detail criteria using in the course USN PRO1000

CRITERION	DEFINITIONS OF RUBRICS' SCALES		
	BEGINNING	*DEVELOPING*	*ACCOMPLISHED* *EXEMPLARY*
2.4. To what extent the software development approaches are adopted?			
2.5. To what extent are management tools used?			
2.6. To what extent are the project role fulfilled?			
2.7. What is the customer's perception on the team work?			
3. The quality of final prototype			
3.1. How much of the products are complete according to the initial requirements?			
3.2. What does the supervisor think about the final prototype?			
3.3. What does the customer think about the final prototype?			

TABLE 2.4   An example of how final evaluation is performed

GROUP	FINAL REPORT (%)	TEAMWORK AND PROCESS (%)	PROTOTYPE (%)	FINAL EVALUATION (%)
Team 1	80	90	100	90
Team 2	70	65	80	71.7

The point aggregation is calculated as a percentage, illustrated by Table 2.4. Fifty percent are passable performance, and 100% are excellent performance. The final percentage will be transformed into a standard grade.

Tip 4: Aggregation of evaluations should be done at the general criteria level.

2.5 UNCERTAINTIES WHEN PLANNING THE COURSE

There are always uncertainties when planning for the course. This section discusses these challenges from the course responsible perspective.

2.5.1 Task clarity

Task/requirement clarity describes how clearly requirements are presented to and understood among project stakeholders in the early phases of projects. Ambiguous requirements or tasks often lead to different interpretations among students that lead to mistakes and confusion and consequently wasted efforts. For non-experienced customers who are not familiar with the course (a researcher or a non-IT officer), it is often difficult to understand what customers really want in the beginning. For technical customers who have in-depth knowledge about the problem space, there can be another risk that they describe the task too detailed that are not easy for new developers to understand.

From our experience, the initial problem descriptions giving to students vary in term of both detail and precision. It is often that the customers come up with a rough idea about what they want to achieve with the project. Therefore, the first meeting with the customer is very important to understand the task requirements. Customers are often busy, and their time is expensive (think about company charge for consulting hour). The teacher needs to plan how to organize a meeting to introduce all customers and then coordinate group meetings between the customers with their student groups. The students need to plan how to most effectively utilize the meeting time with their customers.

Tip 5: The first meeting with customer should be very well planned. Students should meet their customer peer group in the meeting.

2.5.2 Requirement change

It is known that customers are often not aware of what they need in the beginning of the project. Requirement engineering is the process dealing with the change of the requirements. For an actual project, it is quite important for project managers to understand how likely a requirement will change over time. Requirement changes have a significant impact on project outcome when it occurs later in the project life cycle. In traditional projects, implementing the requirement changes in later phases causes 200 times more cost than implementing the requirement changes in analysis phase.

For students, customers who have not experienced with student work might think of them as professional services. Students would probably not be able to handle requirements, as professional developers would do. Experience would help them to quickly understand the scope of the project, freeze the list of requirements for each Sprint, and set up negotiation points when there are significant changes. While requirement changes are an inevitable part of the project, adopting agile practices would help to manage the changes in the student project context.

Tip 6: Early guidance to adopt agile processes and practices would help to deal with requirement changes.

2.5.3 Project difficulties

As mentioned in Section 2.1.2, the difficulty of a project comes from how difficult it is to transform the project uncertainty to a certain domain and how difficult it is to implement the tasks in the certain domain. There is a wide range of tasks in a project course, for instance, from web development, embedded system prototyping, to artificial intelligent algorithms. Task might include also market research and domain study. It is not difficult to implement a project requiring fundamental programming. However, it is difficult to deal with new technologies, especially ones building on top of advanced algorithms, frameworks, or software. The perception of project difficulty is situational depending on students' background and competence. For instance, a mobile app project might not be difficult for fourth-year students with background on android programming, database, and web design. However, it is challenging for first students who learn on HTML programming.

From the teacher's perspective, the project should be set in a manageable difficulty if possible. Selecting a project requires a consideration of current background and knowledge of students. A persona with average competence can provide a benchmark for analyzing the manageable range. It is best that teachers or supervisors have the capacity to implement fully or partly the problems. Uncertainty of solution implementation will be manageable, and teachers/supervisors can support the students with technical details if needed.

> Tip 7: Setting a range of manageable difficulties that teachers or supervisors can provide technical supports.

2.5.4 Customer involvement

The presence of a real customer is important to reflect the real world, some realistic challenges such as requirement changes, and task clarity would be hard to replicate in a simulated context. As defined in Section 2.1.1, customers are expected to actively be involved in the student projects from the beginning until the end of the project. Typically, customer will commit to help students to understand the actual need of the desired products and to give early feedbacks on prototypes. Customers do not need to be full-time present, but available for scheduled meetings. Besides, some customers are actual developers so their technical know-how can directly benefit student teams.

Depending on the customers, students might be required to collaborate in the way that they need to improve their technical knowledge, working process, and professional attitudes. In some cases, students might participate in ongoing real projects in the customer's organization. The student's deliveries can be fed directly into the mainstream of some products. The students might also need to meet the customers of the customer, who are the actual users of a general system.

There are also cases that the involvement of customers is insufficient, where much uncertainties remain in the problem domain, and consequently, students are late to move into solution domains. In such a situation, the role of teacher or supervisor is to timely intervene, in some case to act on the behalf of the customers, to make sure students can move on to the next phase of learning in their project.

> Tip 8: Students should understand the level of engagement and the expectation of collaboration from their customers in early stage of the project.

2.5.5 Experience with methodology and technology

Lacking technical know-how is a global challenge for all kinds of software development teams, whether they are professional or student groups. Obtaining know-how on a specific task reduces the learning time of the team. Goodman and Leyden argued that task familiarity has a positive effect on task performance because every task requires unique configurations of machinery, physical environment, and work activities; therefore, team members' specific knowledge about these aspects of their work can make them more productive (Goodman and Leyden 1991). Similarly, experience with working methods would likely reduce team learning time. As students are required to follow a software development method, in most of the cases are Scrum, they need to perform actual practices and reflect on the adoption of these practices. Teams with one or more members' familiar with Scrum can reach a stable state of productivity faster.

> Tip 9: Both prior knowledge about technologies and process are significant advantage for a team to work on their project.

2.5.6 Free-rider

"Free-riding" in teamwork occurs when one or several members of a group contribute so little to a group project that if the same grade is given to all members of the group, the grade would be misleading and unfair. In our course, we have an explicit focus on the teamwork aspect, so the students are evaluated against how they behave as a team member and how they contribute to the group project. Even for a general project course where individual learning outcome is highlighted, learning only by studying other's work would indicate a superficial learning strategy rather than a deep learning strategy. In any case, free-riding is the unwanted situation and needs to be addressed.

There are different levels when considering a free-riding case. At the first level, a student might pretend that they have done their job and shallowly report their parts in the group delivery. In this case, students participate in all the tasks they are assigned, but it often lacks quality and course extra work to other group members. At the second level, a student might have relatively less commitment to common tasks, miss group meetings, lack of interactions with other team members, hence, do not fulfill their responsibilities in the team. At the third level, the student rarely participates in common activities, does not response to other team members. He or she might be absent in a long period without any compensation for their missing parts in the common projects. Rarely but what happens is that a student who drops out of the course might still be included in a group report.

Dealing with free-riding will be seen at first as an internal team process. First and foremost, it is the team's duty to find out why the situation is happening and not assume that a student is being lazy or is at fault. This might simply mean that the team might not be organized well. Understanding the nature of the contribution problem can help identify the right solution, to keep the concern still as an internal process, or an academic intervention. From project management perspective, leadership and coordination should be performed to quickly detect the situation and provide appropriate counter-measures. The team often has some forms of team agreement in the beginning of the project, works as an informal team contract, and gives them possibility to apply hard mechanism to deal with unwanted situations. When internal attempts are ineffective, the case can be brought to the team supervisor or the teacher of the course, who can intervene in course mechanism. The free-riding issue needs to be identified as early as possible in the project. Waiting until the end of the project would mean overlooking serious team dysfunction and a student that, for one of several reasons, isn't making an adequate contribution.

Tip 10: It is necessary to detect the free-riding situation as early as possible and perform appropriate countermeasures at a team or at the course level.

Project
planning

3

This chapter covers different topics of project planning and some suggestions and ideas from students' perspective. This should be treated as "food for thought," not a step-by-step instruction to perform such planning. Particularly, the chapter discusses about scope planning (Section 3.2), time planning (Section 3.3), risk planning (Section 3.4), communication planning (Section 3.5), way of working (Section 3.6), and usage of tools and infrastructures (Section 3.7). The chapter is particularly written for students who are about or under a software development project.

3.1 INTRODUCTION TO PROJECT PLANNING

The project management plan defines how the project is executed, monitored and controlled, and closed. The project management plan's content varies depending on the application area and complexity of the project. Knowledge about project management plan is well presented in books about project management, such as PMBOK (PMBOK 2013). In the scope of this book, we do not aim at providing details of every project management knowledge area. The content in this section should be seen as tips to solve most common challenges when students conduct their project management. As described in Figure 3.1, seven topics about project planning will be presented.

Project planning is widely thought to be an important contributor to project success. Many benefits of project planning have been proved in different project contexts, including clarification of project uncertainty, reduction of project risks, keeping track of deadlines, delivery of the agreed products. To teach the students about the importance of project planning might not be a simple task. In the beginning, students would probably do not take the lessons through theory lectures. The importance of planning will be better perceived when the students make some mistakes during their projects.

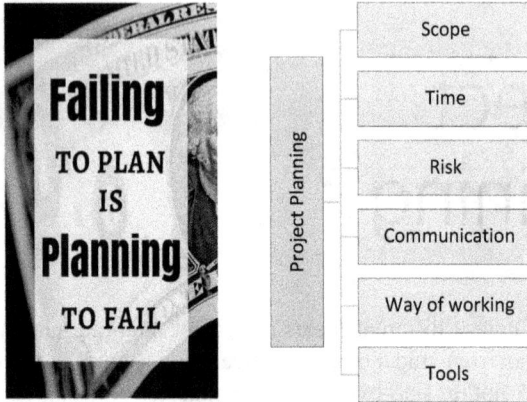

FIGURE 3.1   Topics of student project planning.

Planning is an iterative effort. Project planning can only be based on available information. At the beginning of a project, there are many uncertainties in the available information, and some information about the project and the product may not be available. As the project develops, more and more information becomes available and uncertainties are resolved. The project plan therefore must be reviewed and updated regularly to reflect this changing information environment. The subsections below present important notes for planning a student project work.

Tip 1: Students should do planning early, accepting imperfection and updating their plan iteratively

3.2 SCOPE PLANNING

Scope is the defined features and functions of a product or the scope of work needed to finish a project. There are two ways to look at a project scope:

- Artifact-based scope: for instance, a product composes of three modules. Once the three modules are delivered, the scope of the project is fully covered
- Task-based scope: for instance, a project includes three activities that need to be complete.

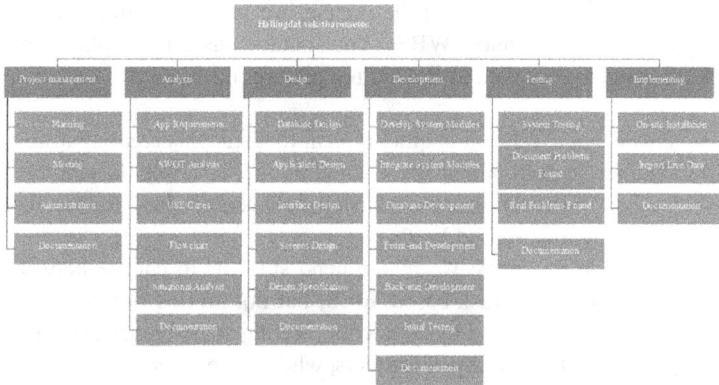

FIGURE 3.2   An example of a task-based WBS.

Correspondingly, there are two ways of defining a work breakdown structure for a project. A work breakdown structure (WBS)[1] is defined as a hierarchical decomposition of the total scope of work to be carried out by the project team to accomplish the project objectives and create the required deliverables (PMBOK 2013). WBS provides the basis for planning and managing project schedules, costs, resources, and changes (Schwalbe 2019, p. 213) There are several approaches for developing a WBS, including using guidelines, the analogy approach, the top-down approach, the bottom-up approach, and mind mapping (Figure 3.2) (Schwalbe 2019, p. 233).

Not every breakdown of project deliverable can be classified as a WBS. WBS has certain characteristics:

- Hierarchy: The WBS is hierarchical in nature. Each "child" level exists in a strict hierarchical relationship with the parent level. The sum of all the child elements should give you the parent element.
- 100% rule: Every level of decomposition must make up 100% of the parent level. It should also have at least two child elements.
- Mutually exclusive: All elements at a particular level in a WBS must be mutually exclusive. There must be no overlap in either their deliverables or their work. This is meant to reduce miscommunication and duplicate work.
- Outcome-focused: The WBS must focus on the result of work, i.e., deliverables, rather than the activities necessary to get there. Every element should be described via nouns, not verbs. This is a big source of confusion for beginners to WBS.

---

[1] https://en.wikipedia.org/wiki/Work_breakdown_structure

During many years of teaching the course, we observe common mistakes when students learn to make WBSs. You should consider the following as a checklist to make sure your WBSs do not picture this:

- This is not a WBS: it is quite typical that students creating a website map instead of a WBS when planning for a web development project. There is also a confusion between product architecture with an artifact-based WBS.
- Mix of WBS types: we recommend students to choose between artifact-based or task-based WBS, but not mixing both of them. It has been observed that students creating some branches of their WBS with a hierarchy of artifacts, while some other branches with a hierarchy of tasks. It is difficult to do time planning, task assignment when referring to this WBS.
- Violation of 100% rule: most of first attempts in creating WBS by students violate the 100% rules. For instance, when thinking about the tasks they need to perform, the students often forget about project planning and management. In fact, they spend a significant amount of time for these tasks. Or students also spend a lot of time for learning new technologies. But when they think about the structure of their products, they do not make a room to include the learning activity in the WBS. Consequently, hundreds of team working hours are not reflected in their WBS and in their time plan.
- Violation of the mutual exclusive rule: there are WBSs with overlapping nodes. For instance, one puts usability testing as a part of the Testing node, as well as a part of project deployment node.
- Right level of details: some WBSs have only one level down, while some others have four, five levels down. Not every node needs to be broken down to the same number of levels. However, the level of details should be determined to keep the balance between the overview of the project and the detail of the tasks.
- Numbering WBS's nodes: it is a common mistake that students do not number their WBS. After creation of the WBS, levels should be numbered for ease of location. For example, a project is divided into two major deliverables, and these are numbered as 1 and 2, respectively. Then, work packages under each major deliverable are numbered as 1.1, 1.2, etc., for major deliverable 1 and 2.1, 2.2, etc., for major deliverable 2.
- Forget to update WBS: as the knowledge about the project evolves, there is a need to update your WBS to reflect correctly the current scope of the project. Maintaining an accurate version of WBS will provide a consistent reference when discussing, assigning tasks and to update time and project plans.

Tip 2: In general, it is preferable to look at task-based WBS with three levels of details

3.3 TIME PLANNING

The essential of time planning is to create and update a project schedule diagram called Gantt Chart. A Gantt chart, or harmonogram, is a type of bar chart that illustrates a project schedule.[2] This chart lists the tasks to be performed on the vertical axis and time intervals on the horizontal axis. The width of the horizontal bars in the graph shows the duration of each activity. Gantt charts illustrate the start and finish dates of the terminal elements and summary elements of a project. Terminal elements and summary elements constitute the WBS of the project. Modern Gantt charts also show the dependency (i.e., precedence network) relationships between activities.

Similarly to Section 3.2, we present here common mistakes when students firstly making Gantt charts:

- This is not a Gantt chart: there are common conventions about drawing Gantt chart. Students should use existing templates for visualizing their Gantt Chart and not invent their own notations. Microsoft Office provides a template as an Excel document,[3] but you can find various templates in other software online.
- Right size of Gantt chart: as seen in Figure 3.3, the overview of the whole project is captured in one screen. It is possible to go in details of the WBS and have the correspondingly timeline for each task. However, keep in mind that too many details can make it difficult to watch the overview. Some modern project management tools might provide features to dynamically switch between views of Gantt Chart. However, most of the free templates will not give you the capacity.
- Forget to add milestones: Gantt chart includes not only timeline for activities, but also important milestones, such as Sprint review meeting, User acceptance testing, Project final presentation, and Final exam date. Visualizing milestones will guide the adjustment of time plan to meet project deadlines.

---

[2] https://en.wikipedia.org/wiki/Gantt_chart
[3] https://templates.office.com/en-us/simple-gantt-chart-tm16400962

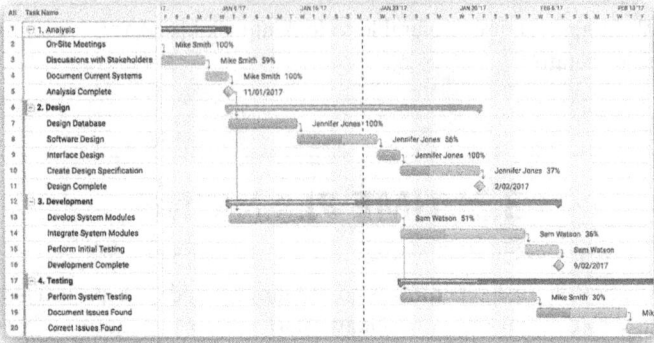

FIGURE 3.3   An example of a Gantt chart for a web application development project.

- Forget to update Gantt chart: similar to WBS, time plan can change over time. There is also a need to update the schedule chart when WBS is updated. Maintaining an accurate version of WBS is important to have a correct view on the project progress and the time plan till the end of the project.

Tip 3: Students should use the Excel template for Gantt Chart provided in this book for their time plan

3.4 RISK PLANNING

Risk is defined as a possibility of loss, the loss itself, or any characteristic, object, or action that is associated with that possibility. In a software project, risk implies *"the impact to the project which could be in the form of diminished quality of the end product, increased costs, delayed completion, loss of market share, or failure"* (Boehm 1991). As a part of project management disciplines, risk management activities have been explored in the connection to the success of software projects (Boehm 1991). The management of risk, namely the risk of failing to meet one's expectations within given constraints in budget and/or time, is of paramount importance in every human activity. Project managers should be able to early identify different risks, evaluate their impact, and come up with risk management strategies. Risk planning is an important learning area for students. The main activities in risk management include:

- Risk identification: before risks can be managed, they must be identified. This produces lists of the project-specific risk items likely to compromise a project's success.
- Risk analysis: assesses the loss probability and loss magnitude for each identified risk item, and it assesses compound risks in risk–item interactions, prioritization of risk items basing on their frequency and severity. This includes risk prioritization: produces a ranked ordering of the risk items identified and analyzed.
- Risk resolution: preparing actions when the risk actually happens to produce a situation in which the risk items are eliminated or otherwise resolved.
- Risk monitoring: involves tracking the project's progress toward resolving its risk items and taking corrective action.

Risk can also be categorized due to its nature. Known risks are the risks that can be found after assessing the project plan, the commercial and technological environment, and other reliable informational sources carefully (for example: unrealistic delivery time, no demand, software file, and abominable developing environment). Unknown risks are something perhaps will appear really, but it is very difficult to recognize them coming in advance.

According to the type of tasks, a software project may encounter various types of risks:

- Technical risks include problems with languages, project size, project functionality, platforms, methods, standards, or processes. These risks may result from excessive constraints, lack of experience, poorly defined parameters, or dependencies on organizations outside the direct control of the project team.
- Management risks include lack of planning, lack of management experience and training, communications problems, organizational issues, lack of authority, and control problems.
- Financial risks include cash flow, capital and budgetary issues, and return on investment constraints.
- Contractual and legal risks include changing requirements, market-driven schedules, health and safety issues, government regulation, and product warranty issues.
- Personnel risks include staffing lags, experience and training problems, ethical and moral issues, staff conflicts, and productivity issues.

- Other resource risks include unavailability or late delivery of equipment and supplies, inadequate tools, inadequate facilities, distributed locations, unavailability of computer resources, and slow response times.

If students do not have a practical project course at this scale before, there might be more uncertainties associated with the project. Different from professional projects, student projects will have more concerns regarding the academic context, i.e., course calendar, availability of team member for the project course, personal knowledge, and skills. Table 3.1 describes common risk items that are identified by many student groups over the years.

TABLE 3.1 An example of risk identification and assessment table

NO.	DESCRIPTION	PROBABILITY	SEVERITY	RISK SCORE
R1	Underestimating workload	3	5	15
R2	Lack of programming skills to implement the product	3	3	9
R3	Schedule conflict with other courses	4	5	20
R4	Bad communication inside the team	3	4	12
R5	Lack of common level of commitment to the project	4	4	16
R6	Team member absence for a long time	1	5	5
R7	Customer do not response timely	1	4	4
R8	Required technologies are not accessible			
R9	Misunderstandings of project requirements	5	3	15
R10	Member no longer available	2	5	10
R11	Time to learn new technologies	4	2	8
R12	Uneven distribution of workloads	5	2	10
R13	Poor choice of development methodology	1	5	5
R14	Poor choice of programming framework	2	5	10
R15	Poor leadership	3	3	9

After identifying and analyzing the risks, students should come up with strategies to avoid or to mitigate bad consequence when the risks happen. The strategy should base on a realistic plan considering the capacities of each and all team members.

Tip 4: Students can learn from previous or similar projects about possible risks and work out a resolution plan to deal with them

3.5 COMMUNICATION PLANNING

Formal communication plan is often a written document that guides project communication. The document typically includes information about stakeholder communications requirements, information to be communicated, involving stakeholders, communication methods, frequency of communication, and procedures for resolving issues (Schwalbe et al. 2013).

In a student project, there are typically course teacher, supervisor, student team, and customers. Customers might be a hierarchical group of people from different functional units, but the expected communication with them is often limited. There is often an assumption that stakeholders are communicated in common ways. However, each group has different communication needs. Creating some sort of communications management plan and reviewing it with project stakeholders early in a project helps prevent or reduce later communication problems. A plan can be kept as simple as an overview shown in Table 3.2. As described, the team follows Agile methodology with different Scrum meetings. They also need to communicate with supervisors, course lecturers, and customers either in physical meetings or in online meetings.

It is also important to decide the best way to distribute the information (Schwalbe et al. 2013). Is it sufficient to send written reports for project information? Are meetings alone effective in distributing project information? Are meetings and written communications both required for project information? What is the best way to distribute information to virtual team members? Figure 3.4 gives students an idea about the rank of communication channels basing on their effectiveness. Face-to-face meeting with visual aids, such as blackboard, is the richest and most effective communication channel. Paper-based, i.e., sending posts, is the least effective and also coldest way to communicate.

Many people think they can just add more developers to a project that is falling behind schedule. Unfortunately, this approach often causes more

TABLE 3.2   A communication plan for a project course

COMMUNICATION TYPE	TARGET/ AUDIENCE	REQUIRES	FREQUENCY	MEAN
Team supervision	Project team Supervisors	Report on regular basic, critical, challenges, feedback	Once per week	Physical meeting
Standup meeting	Project team	Report on daily work, challenges, and feedback	Twice per week	Slack
Sprint planning	Project team	What to do in the sprint task assignment	Beginning of Sprint	Zoom
Technical talk	Project team	Discuss technically among relevant team members	Daily	Slack
Project postmortem	Project team	Analyzing the project and determining the project success or failure	When the project is done	Slack
Usability testing	Test users End user	Testing UX of the product	Final sprint	Physical meeting
Sprint review	Project team customer	Project progress, feedback from the community, potential end user's consideration	End of sprint	Zoom
Final demo	Project team Customer Supervisors	Present the delivery of the project, lesson learned	End of the course	Physical meeting

setbacks because of the increased complexity of communications. People are not interchangeable parts. It is not equivalent between a task of 2 months working by one person and a task doing by two people in 1 month.

Rarely does the receiver interpret a message exactly as the sender intended. Therefore, it is important to provide many methods of communication and an environment that promotes open conversation.

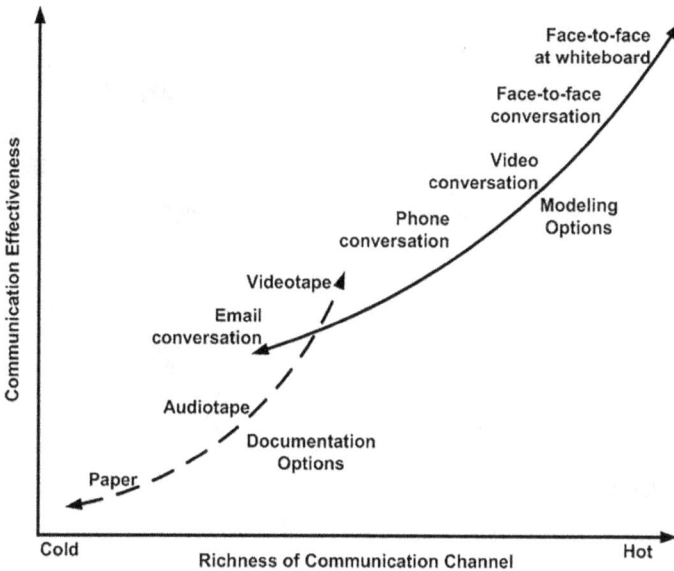

FIGURE 3.4 The order of common communication channels.

> Tip 5: Each communication channel will fit into different communication context.

3.6 WAY OF WORKING

Software engineering presents a vast amount of software development models, processes, and practices. According to Jacobson, the author of Rational Unified Process, there is no one-size-fits-all when it comes to methods (Jacobson et al. 2019). A method needs to be adapted to the project situation, which is inherently evolving. Project teams need to constantly evolve their methods as long as there is work to do on the product.

There are some other software development approaches worth mentioning, such as waterfall life cycle, spiral life cycle model, incremental software development, RAD life cycle model, that occur early in the history of software industry. In the last 20 years, methods such as Scrum and Disciplined Agile are among the most common approaches for small software development team. Ken Schwaber and Jeff Sutherland are the ones who formulated

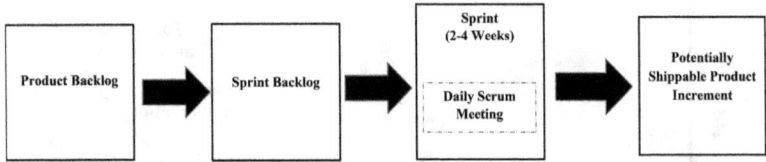

FIGURE 3.5   A typical Sprint in a Scrum process (Scrum Guide).

the initial versions of the Scrum framework and presented Scrum as a formal process at the conference OOPSLA'95 (Sutherland et al. 1997). In 2001, a group of 17 software professionals (including Ken and Jeff) published the Manifesto for Agile Software Development[4] that focuses on four values:

- Individuals and interactions over processes and tools
- Working software over comprehensive documentation
- Customer collaboration over contract negotiation
- Responding to change over following a plan

Scrum is an Agile process that allows us to focus on delivering the highest business value in the shortest time. It allows us to rapidly and repeatedly inspect actual working software (every 2–4 weeks). The business sets the priorities. Teams self-organize to determine the best way to deliver the highest priority features. Every 2 weeks to a month anyone can see real working software and decide to release it as is or continue to enhance it for another sprint, as seen in Figure 3.5.

In subsections below, we present the list of Scrum practices that we think most relevant for the student project contexts. We present briefly the practice, and students who are going to adopt them can find more instructive guidelines in other sources. The full list of practices can be found in the Industrial guideline.[5]

Tip 6: Students should have their own way of working by composing existing practices in the way that works best for their project situation.

3.6.1 Product backlog and prioritization

A list of everything that needs to be done to the product is listed in priority order. The sprint backlog is a list of tasks identified by the Scrum team to be completed during the Scrum sprint.

---

[4] https://agilemanifesto.org/
[5] https://www.scrumguides.org/scrum-guide.html#purpose

3.6.2 User stories

User stories are short, simple descriptions of a feature told from the perspective of the person who desires the new capability, usually a user or customer of the system. They typically follow a simple template:

As a < type of user >, I want < some goal > so that < some reason >.

User stories are often written on index cards or sticky notes, stored in a shoe box, and arranged on walls or tables to facilitate planning and discussion. As such, they strongly shift the focus from writing about features to discussing them. In fact, these discussions are more important than whatever text is written.

3.6.3 Sprint plan meeting

Creating a sprint plan is a process that the scrum team needs to discuss together to review each of the tasks in the breakdown. Sprint Planning answers the following:

- What can be delivered in the Increment resulting from the upcoming Sprint?
- How will the work needed to deliver the Increment be achieved?

3.6.4 Daily standup meeting

As the sprint begins, it is always important to have a call on a daily basis so that everyone on the team is aware of any issues or concerns and that the task is in progress. Below are three questions to ask or answer during this daily call:

- What did you do today?
- What will you do tomorrow?
- Is anything holding up your progress (impediments)?

Daily standup meeting needs to be adjusted to fit into course context. Students can conduct virtual standup meeting via video call, for instance. Students can also have a biweekly standup meeting instead of doing the meeting every day.

3.6.5 Retrospective meeting

Although a good Scrum team will be constantly looking for improvement opportunities, the team should set aside a brief, dedicated period at the end of each sprint to deliberately reflect on how they are doing and to find ways to improve. This occurs during the sprint retrospective. The Scrum Master asks each team member to identify specific things that the team should:

- start doing
- stop doing
- continue doing

3.6.6 Sprint review meeting

During this meeting, the team shows what they accomplished during the sprint. Typically this takes the form of a demo of the new features. The sprint review meeting is intentionally kept very informal, typically with rules forbidding the use of PowerPoint slides and allowing no more than 2 hours of preparation time for the meeting. In the context of the course, students might organize a few Sprint review meetings depending on the availability of their customers.

3.6.7 Planning poker

Planning Poker[6] is a common approach to estimate the size of items in product backlogs. Planning Poker can be used with story points, working hours, or any other estimating unit. To start a poker planning session, the product owner or customer reads an agile user story or describes a feature to the estimators. Each estimator is holding a deck of Planning Poker cards with values such as 0, 1, 2, 3, 5, 8, 13, 20, 40, and 100, which is the sequence we recommend. The values represent the number of story points, ideal days, or other units in which the team estimates. The estimators discuss the feature, asking questions of the product owner as needed. When the feature has been fully discussed, each estimator privately selects one card to represent his or her estimate. All cards are then revealed at the same time. If all estimators selected the same value, that becomes the estimate. If not, the estimators discuss their estimates. The high and low estimators should especially share

---

[6] https://en.wikipedia.org/wiki/Planning_poker

their reasons. After further discussion, each estimator reselects an estimate card, and all cards are again revealed at the same time. The poker planning process is repeated until consensus is achieved or until the estimators decide that agile estimating and planning of a particular item needs to be deferred until additional information can be acquired. Students should learn to use planning poker for every Sprint. However, it is noticed that for novices, project estimates could be overoptimistic and planning poker additionally increased the overoptimism (Mahnič and Hovelja 2012).

3.6.8 Code standard

If programmers all adhere to a single agile coding standard (including everything from tabs vs. spaces and curly bracket placement to naming conventions for things such as classes, methods, and interfaces), everything just works better. It's easier to maintain and extend code, to refactor it, and to reconcile integration conflicts, if a common standard is applied consistently throughout.

3.6.9 100% coverage for unit tests

Unit tests are released into the code repository along with the code they test. Code without tests may not be released. A program with high test coverage, measured as a percentage, has had more of its source code executed during testing, which suggests that it has a lower chance of containing undetected software bugs compared to a program with low test coverage

3.6.10 User acceptance test

Acceptance tests[7] are created from user stories. The customer specifies scenarios to test when a user story has been correctly implemented. A story can have one or many acceptance tests, whatever it takes to ensure that the functionality works.

Acceptance tests are black box system tests. Each acceptance test represents some expected result from the system. Customers are responsible for verifying the correctness of the acceptance tests and reviewing test scores to decide which failed tests are of highest priority. Acceptance tests are also used as regression tests prior to a production release.

---

[7] https://en.wikipedia.org/wiki/Acceptance_testing

A user story is not considered complete until it has passed its acceptance tests. This means that new acceptance tests must be created each iteration, or the development team will report zero progress.

3.6.11 Burndown chart

Early detection of issues is always better than late detection. This can be done as long as progress is tracked. The daily burndown chart[8] is a simple practice that helps the team track daily progress. Scrum methodology uses the burndown chart to see the progress on completed and pending tasks.

3.6.12 Pair programming

Pair programming[9] is a commonly adopted Agile technique in which two programmers work together at one workstation. One, the driver, writes code while the other, the observer or navigator, reviews each line of code as it is typed in. The two programmers switch roles frequently.

3.6.13 Test-driven development

Test-driven development (TDD)[10] starts by writing a test for a user story or a feature. The first step is to quickly add a test, basically just enough code to fail. Next you run your tests, often the complete test suite although for the sake of speed you may decide to run only a subset, to ensure that the new test does in fact fail. You then update your functional code to make it pass the new tests. The fourth step is to run your tests again. If they fail, you need to update your functional code and retest. Once the tests pass, the next step is to start over.

3.6.14 Scrum master

Development process and practices facilitated by a dedicated role. The Scrum Master[11] is responsible for making sure a Scrum team lives by the values and practices of Scrum. The Scrum Master is often considered a coach for the team, helping the team do the best work it possibly can. The Scrum Master

---

[8] https://en.wikipedia.org/wiki/Burn_down_chart
[9] https://en.wikipedia.org/wiki/Pair_programming
[10] https://en.wikipedia.org/wiki/Test-driven_development
[11] https://www.scrum.org/resources/what-is-a-scrum-master

can also be thought of as a process owner for the team, creating a balance with the project's key stakeholder, who is referred to as the product owner. The Scrum Master does anything possible to help the team perform at their highest level. This involves removing any impediments to progress, facilitating meetings, and doing things such as working with the product owner to make sure the product backlog is in good shape and ready for the next sprint. The Scrum Master's role is commonly filled by a former project manager or a technical team leader but can be anyone.

The Scrum Master is also often viewed as a protector of the team. The most common example is that the Scrum Master protects the team by making sure they do not overcommit themselves to what they can achieve during a sprint due to pressure from an overly aggressive product owner. However, a good Scrum Master also protects the team from complacency.

3.6.15 Product owner

The product owner[12] is typically a project's key stakeholder. Part of the product owner responsibilities is to have a vision of what he or she wishes to build and convey that vision to the scrum team. This is key to successfully starting any agile software development project. The agile product owner does this in part through the product backlog, which is a prioritized features list for the product.

The product owner is commonly a lead user of the system or someone from marketing, product management, or anyone with a solid understanding of users, the marketplace, the competition, and future trends for the domain or type of system being developed.

3.7 TOOLS AND INFRASTRUCTURE

The recent surge in home-based workers due to the disease COVID-19[13] has made it a necessity to use tools that allow students to collaborate from afar. Fortunately, students have an abundance of collaborative tools and approaches that they can choose from when collaborating with others. A collaborative tool helps people collaborate. These tools differ in terms of their functionality, usability, target audience, etc. An understanding of the suitability of different tools for different purposes would allow us to provide students with the most suitable tools for each learning scenario. Besides, there are also

[12] https://www.scrum.org/resources/what-is-a-product-owner
[13] https://en.wikipedia.org/wiki/Coronavirus_disease_2019

development tools for the team to design, develop, and test their products, to coordinate meeting and to manage the project progress, as described in Table 3.3. We also experience that students' preferred toolset varies almost year to year (Figure 3.6).

Tip 7: Students should consult course teachers and supervisors about a suitable set of collaborative and development tools

TABLE 3.3 Categories of tools for a software development project

TYPE	DESCRIPTION	EXAMPLES
Communication	Provide exchange of information between individuals	Gmail, Outlook, Facebook Messenger, Slack, Discord, etc.
Coordination	Allow a person to set up group activities, schedules, and deliverables	Trello, Google doc, Outlook Calendar, Jira, Team, etc.
Cooperation	Have real-time discussions and to shape an idea or thought together	Skype, Zoom, appear.in, whereby.com, Team, etc. MockFlow, Adobe Xd, Draw.io, etc.
Text and multimedia editor	Create texts, diagrams, figures, audio, videos, etc.	Google doc, Sublime, Visual Paradigm, Lucichart, Draw.io, Sketch
Common workspace	Cloud-based storages of project materials	Google Drive, One Drive, Dropbox, Team, etc.
Code editor	Write, compile, run, and debug code	Sublime, .Net framework, Atom, Bracket, Eclipse, etc.
Code repository	Store and manage versions of source code	GitHub, Bitbucket, GitLab
Application framework	Ready-made libraries, frameworks for certain types of software	jQuery, node.js, react native, Angular, etc.
Webserver	A server software to host back-end applications	XAMPP, Apache, Nginx, Microsoft IIS, etc.
Database	Collections of data, can be relational or non-relational databases	MySQL, MongoDB, SQL server, Firebase, PostgreDB, etc.

FIGURE 3.6   The toolset in project course at USN, Spring 2020.

Before students can take the best from using tools in their projects, they should be aware of several common challenges we have observed during several years of the project courses:

- Collaborative attitudes: before tools are put in use, the collaborative mindset should be established. There is no use to adopt different technologies to bridge the gap of process, geographical, or temporal distances while people are not willing to use them.
- Use a lot of tools at the same time: There are so many digital tools available for a teamwork to choose from today – many of which do the exact same thing. It might end up that everyone in the team putting their preferred tools on the table, many are chosen, and they both serve for the same communication channels.
- Learn to use tools: while some tools are straightforward to use, or students have already decent experience with them, many tools provide advanced features that boost productivity and team performance. These features need to be discovered by practices. We observe that many students do not use Trello,[14] a rather simple tool, till the end of their projects. Some other tools, such as Jira,[15] GitHub,[16] or Docker,[17] are not effectively adopted most of the time.

---

[14] https://trello.com/
[15] https://www.atlassian.com/software/jira
[16] https://github.com/
[17] https://www.docker.com/

- Tools do not solve your problems: many students choose to explore complicate development environment, such as Visual Studio[18] or Xamarin,[19] to develop a website that can be efficiently written with Sublime Text.[20] Tools might provide some utility for you to conduct your work faster. However, especially with in the academic context, it is not the focus on using complex tools. The roles of tools in a software development project need to be communicated properly to students in the beginning of the course.

---

[18] https://visualstudio.microsoft.com/
[19] https://dotnet.microsoft.com/apps/xamarin
[20] https://www.sublimetext.com/

Project execution

4

This chapter covers different topics of project execution, theories, and specific examples. While the chapter mainly provides theoretical foundations, it can also be referred to for some detailed instructions when students are carrying out their project. Particular, the chapter addresses the concepts about problem space and solution space (Section 4.1), a process of teamwork development (Section 4.2), process monitoring (Section 4.3), and most importantly, different aspects of Software Engineering (Section 4.4).

4.1 PROBLEM SPACE VS. SOLUTION SPACE

There is a pedagogic question whether engineering education curriculum focuses enough on giving exercises to students that force them to confront underdetermined and ill-defined problems (Schraw et al. 1995)? In many programs, senior students did better than freshmen at less creative or more typical design problems, which were already well defined for them, when the solution generation and evaluation process was linear and straightforward. In contrast, when faced with an open-ended, less-defined design problem, younger students did better.

In a project context, customer needs define the problem, which is actually not well defined. Students need to go through a process of identifying and eliciting requirements from customer needs. It should also take into account the constraints that influence the problem space. The problem in software engineering varies in different context. It can be simply an idea of product from the CEO of your company that you and your team need to implement a solution. It can be specified features from a customer for their software. It can also be a need from a segment of market that is not yet addressed.

51

When speaking about solution space, any product or the product design, such as Minimum Viable Product, mock-ups, wire-frame, and prototype (Nguyen-Duc and Abrahamsson 2016, Nguyen-Duc et al. 2020), depends on and is built upon problem space, but is in solution space. So, we can say problem space is at the base of solution space. Solution space includes any product or representation of a product that is used by or intended for use by a customer. When you build a product, you have chosen a specific implementation. Whether you've done so explicitly or not, you've determined how the product looks, what it does, and how it works.

Solution space consists of many possible solutions for a given problem. There is also matter to what extent a choice of solution addresses the given problem. A prototype can partly resolve the need of customers when they see how their problem can be actually solved. A product can realize the solution and put it into operationalization.

In some circumstances, the process of going through a problem space and a solution space does not happen one after another, but in parallel. Problem and solution are both explored in a mutual manner. Nguyen-Duc describes this intertwined process as a hunter-gatherer model. A hunter tasked to find an innovative idea and a gatherer tasked to implement the idea. While hunting the idea though ambiguity spaces has change-driven, analytical, and qualitative nature, gathering the idea across predetermined paths has plan-oriented, optimizable, and quantitative nature (Nguyen-Duc et al. 2015). As an innovation creation process, a typical cycle of hunting starts from generation of many ideas, then coming to a small set of features and exploring of customer and market, as seen in Figure 4.1. A cycle of gathering includes implementing features as prototypes, complete products, and enhanced products in terms of quality, i.e., performance and scalability. Both hunting and gathering patterns can be repeated in a short period (from a day to a month). There is also a transition from the hunting cycle to the gathering cycle when the set of features is determined and sent to the development phase.

Concrete tactics to explore an uncertain space include (1) guessing and checking; (2) making an orderly list; (3) eliminating possibilities;

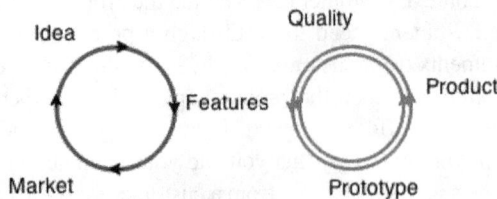

FIGURE 4.1 The design thinking process from problem space to solution space (Nguyen-Duc et al. 2015).

(4) considering special cases; (5) using direct reasoning; (6) looking for a pattern; (7) drawing a picture; (8) solving a simpler problem; (9) using a model; (10) and working backward.

We suggest the students make a short presentation about their understanding about the project, to make sure the discovery through problem domain is guided and supported. A typical presentation should include:

- Your understanding about the project business context
- Present top 3 functional requirements you think are most important for the project
- Present nonfunctional requirements that are important for the project
- Present technologies you think can be used
- Current road-blocking challenges

Students also need to prepare themselves for a steep learning curve in this course. On the one hand, the course is designed to be close to real-world problem, students are expected to be contributors to professional work. On the other hand, students might just equip themselves with fundamental programming experience and probably lack most of practical experience with teamwork, process, and desired knowledge domains, which are often specialized to customer's work. This creates a knowledge gap that is often not closed by lectures or supervision. Students are expected to learn themselves, if needed, new programming language, new software platform, new architecture, new framework, new working process, new business domain, etc. They should also be able to put these knowledges together in a form of deliverable artifacts.

We recommended that students create Sprint 0, or a prestudy phase, where most of effort focuses on explorative activities to understand what are there in the solution space and in the problem space. After that, a learning process should be planned to acquire the necessary knowledge to produce the solution for the understood problem.

4.2 TEAMWORK MONITORING

The **forming–storming–norming–performing** model of group development was first proposed by Tuckman, who said that these phases are all necessary and inevitable in order for the team to grow, face up to challenges, tackle problems, find solutions, plan work, and deliver results (Tuckman 1965).

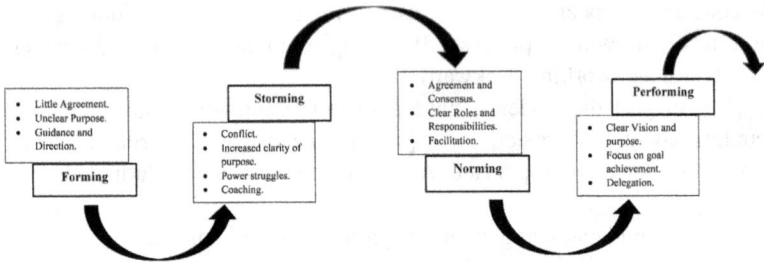

FIGURE 4.2   Forming–storming–norming–performing model of team development (Tuckman and Jensen 1977).

We found that the model applies well to describe the development of student team in our courses. We skip the adjourning stage, which is added later to the model, as it is not much relevant in student project context.

As seen in Figure 4.2, the first stage of team development is Forming. The team meets and learns about the opportunities and challenges and then agrees on goals and begins to tackle the tasks. Team members tend to behave quite independently. They may be motivated but are usually relatively uninformed of the issues and objectives of the team. Team members are usually on their best behavior but very focused on themselves. Mature team members begin to model appropriate behavior even at this early phase. Members attempt to become oriented to the tasks as well as to one another.

Student teams are encouraged to do individual skill assessments – having honest discussions about the individual team members' strengths and weaknesses. Teams do research into their client, their project area, and relevant synthesis and analysis methods. By the end of this stage, the team has developed a plan of activities with milestones and deadlines that are presented in a written proposal.

Tip 1: Students should honestly share their strengths, weaknesses, and motivations toward the course, develop individual and group plans.

The second stage is Storming, where the differences between ideas, work patterns, methods, and behaviors of individuals on the team create conflict. Tolerance of each team member and their differences should be emphasized; without tolerance and patience, the team will fail. This phase can become destructive to the team and will lower motivation if allowed to get out of control. Some teams will never develop past this stage; however, disagreements within the team can make members stronger, more versatile, and able to work more effectively as a team.

The goal for each student team is to get each and every team through the Storming Stage as quickly and efficiently as possible. Students learn that their usual coping methods – such as "I'll do it all myself" or "I can forget about it after the due date" – that worked on projects in "normal" courses don't work due to the magnitude of the projects. Depending on the personalities of the team, the team leader, and the advisor, this Storming stage can be either a mild drizzle or a hurricane.

Tip 2: Students should try to get through the Storming phase quickly.

Norming. This stage happens when the team is aware of competition and they share a common goal. In this stage, all team members take the responsibility and have the ambition to work for the success of the team's goals. They start tolerating the whims and fancies of the other team members. They accept others as they are and make an effort to move on. In a project setting, this is evident as the teams divide the tasks and develop the beginnings of simultaneous rather than sequential activities. The danger here is that members may be so focused on preventing conflict that they are reluctant to share controversial ideas. For example, a team may divide the tasks so that two students work on one task while two others work on another. More creative teams turn the norming process into more of a round-robin activity. Students A and B work on one task; students B and C work on another task; and students A and C work on yet another task.

Tip 3: Task assignment can be optimized in Norming stage

Performing. By this time, they are motivated and knowledgeable. The team members are now competent, autonomous, and able to handle the decision-making process without supervision. Dissent is expected and allowed as long as it is channeled through means acceptable to the team. The team will make most of the necessary decisions. Even the most high-performing teams will revert to earlier stages in certain circumstances. Students come to understand that in order for any of them to be successful, they need to be successful together. Many long-standing teams go through these cycles many times as they react to changing circumstances. For example, a change in leadership may cause the team to revert to *storming* as the new people challenge the existing norms and dynamics of the team (Nguyen-Duc et al. 2019).

4.3 PROCESS MONITORING

Project activities include not only technical and managerial tasks but also administrative tasks. They might not be much mentioned in your team meetings, but they are important to do:

Hour tracking. It is important to keep track of hours worked, both to ensure a fairly distributed workload and to help with future estimations. All members of the team kept a log of hours worked in a spreadsheet, with notes to explain what they were working on. The spreadsheet also has a summary of all members, calculated automatically from each individual sheet. There might not be necessary to keep a precise number of hours every; however, the rough number from each team member should be counted at a week basic or Sprint basic.

> Tip 4: Keep tracking of individual work from Day 1. Tracking can be done in a weekly basic

Meeting. Over the course of a week, a team will have different meetings: internal team meetings, such as Scrum planning or Retrospective meeting, meetings with customers and with supervisors:

- Team meeting: One challenge for scheduling common meetings for all team members is the varying individual schedule. Some teams can have one common meeting per week, some other teams have two to three ones per week. We suggest our students to at least having two meetings per week in the first month of the course, when the forming and storming stage would likely to happen. Besides, we suggest the students use some of the meeting formats according to their development approaches. After some weeks, we expect Scrum meetings, i.e., Scrum planning, daily standup meeting, and Retrospective meeting are followed when the methodology is established. For each meeting, the meeting agenda should be created and sent to all team members beforehand. During the meeting, there should be a meeting secretary who writes meeting notes.
- Customer meeting: Students should actively involve their students into the projects. Depending on the availability of the customers, meeting with them, either physically or digitally should be planned

for at least twice per month. Having this meeting right before the sprint planning allows the team to gather feedback from the customer, which is used when planning the next sprint. Email communication can be an efficient way. Team should assign a contact person who will be responsible for maintaining communication link with the customer, i.e., sending agenda, meeting note, questions, etc.

• Supervisor meeting: An advisor meeting should be held regularly in 1 or 2 week basic. The purpose of this meeting is to keep the supervisor up to date on the team's work, for the team to ask any question around the requirements of the project, get some help, and make sure that the team was working well together. Therefore, teams should be open about their challenges during the meeting, utilize the supervisors as the major source of external support to solve their problems regarding to teamwork, processes, and sometime technical details.

Tip 5: Take note of meetings and store meeting notes online in one place for tracking and simplicity

Task assignment tracking. While students can be flexible on their choices of tools and methods, using Trello as a Kanban board to support Scrum processes is suggested in our course. Trello has a number of advantages, including making progress visible to the whole team and allowing details of every task (such as comments, checklists, due dates, and attachments) to be added to cards. Trello enables all team members to participate in task-oriented discussions, view the workflow, share files and notes, and comment on the various tasks in the workflow. In the beginning of the course, there should be a section to introduce and guide the usage of Trello. We suggest the students to create the following columns:

• Meta column: Information about the team, project, notation used in the team board
• Product backlog column: includes the latest version of the project backlog
• Sprint 1 backlog: includes the list of backlog items planned for this Sprint
• Sprint 1 to-do: when the team starts with this Sprint, items from Sprint backlog shall be copied and pasted into this To-do column
• Sprint 1 complete: when an item is done, it should be moved from Sprint 1 to-do column to the Complete column.

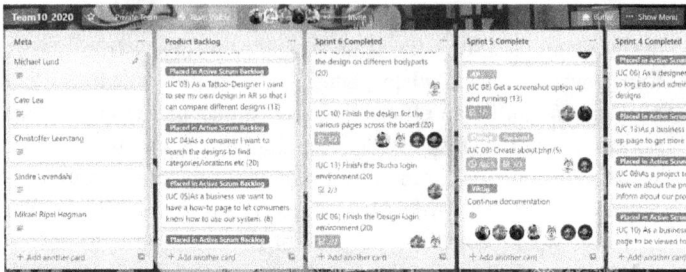

FIGURE 4.3 An example of a Trello board.

Repeating these columns for Sprint 2, 3, and so on, as seen in Figure 4.1. We also encourage students to discover different utility functions in Trello, such as labeling different types of tasks with different colors, adding deadlines to a card, creating checklist for a large task, tagging people responsible for a task, using plugins to Trello that support project management (Figure 4.3).

> Tip 6: Every team member should actively participate in building and maintaining status of tasks in Trello

Cultural difference. During the project progress, students might take care of possible difference among their team members that might come from cultural gaps. A culture gap is any systematic difference between two cultures, which hinders mutual understanding or relations. Such differences include the values, behavior, education, and customs of the respective cultures. For instance, observing in one of our student teams:

> Because our group was a group with two Norwegian and two Serbian students, ours was a group with a potential for language difficulties and misunderstandings due to different cultural backgrounds. This was something we as a group were aware of and discussed in one of our first meetings. Luckily, because all of the members of the group spoke english very well, we did not experience any difficulties due to language.
>
> (*Group6_2017*)

4.4 SOFTWARE ENGINEERING

Software Engineering provides the systematic application of scientific and technological knowledge, methods, and experience to the design, implementation, testing, and documentation of software (ISO/IEC/IEEE 2010).

We present the following Software Engineering knowledge areas, which are the most relevant to student projects: Requirement Engineering (Section 4.4.1), Software Design (Section 4.4.2), Implementation (Section 4.4.3), and Testing (4.4.4). Some other knowledge areas, such as software maintenance, software quality, and software economics, are not covered in the scope of our courses due to the insufficient setting.

4.4.1 Requirement Engineering

In the requirements specification phase, it is important to explicitly state the system requirements and link them to the project description from customers, in a given written document, or via verbal conversation. There are typically two types of requirements, **functional** and **nonfunctional**.

Functional requirements (FR) define what the system does or must not do, Nonfunctional requirements (NFR) specify how the system should do it. FR are usually in the form of "system shall do <requirement>" and NFR are in the form of "system shall be <requirement>". NFR are often called "quality attributes" of a system, for instance:

- Usability. This focuses on how a website appears and how people interact with it. What color are the screens? How big are the buttons?
- Reliability/Availability. What are the uptime requirements? Does it need to function 24/7/365?
- Scalability. As needs grow, can the system handle it? For physical installations, this includes spare hardware or space to install it in the future.
- Performance. How fast does it need to operate?
- Supportability. Is support provided in-house or is remote accessibility for external resources required?
- Security. What are the security requirements, both for the physical installation and from a cyber perspective?

In a Scrum process, requirements are captured in a form of User Story (Section 6.5.2). We suggest students to number their user stories with FR if they describe functional requirement, NFR if the describe nonfunctional requirement, and T if they are tasks. The advantage with numbering is that it is then easy to separate the requirements from the rest of the text, each becomes explicit, and you achieve traceability and structure. An example of user stories:

- FR1: As an admin I would like to add events to a calendar so that everyone can see upcoming events.

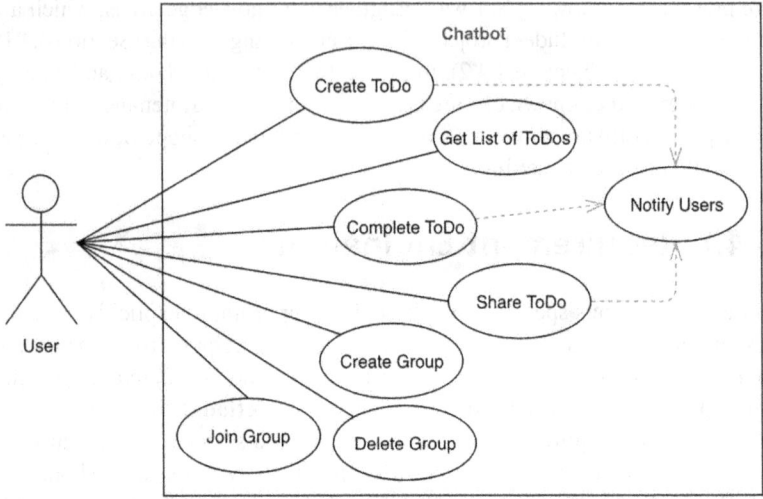

FIGURE 4.4 An example of use-case diagram for a chat-bot application.

- NFR1: As a user, I would like to get to the task board within three clicks since the web is opened.

The detail analysis of the requirements might be a required part of a long report (ca. more than 100 pages). We recommend students also create use-case diagrams here, because we then can make quick and reliable estimates of the ensuing design, programming, and test effort, as shown in Figure 4.4. Some further specification of nonfunctional or quality requirements of the product is also useful.

4.4.2 Design

Architectural patterns. An architectural pattern is a general, reusable solution to a commonly occurring problem in software architecture within a given context (Taylor et al. 2009). Such patterns provide a way of describing good design structures, such that they can be reused in a range of different implementations and can help the team to avoid bad design decisions. Some common patterns include:

- Client-server pattern: consists of a server and multiple clients. The server component will provide services to multiple client components.

- Pipe-filter pattern produces and processes a stream of data. Each processing step is enclosed within a filter component. Data to be processed is passed through pipes.
- Broker pattern: is used to structure distributed systems with decoupled components. These components can interact with each other by remote service invocations. A broker component is responsible for the coordination of communication among components.
- Model-view-controller (MVC) pattern: divides an interactive application into three parts as (1) model – contains the core functionality and data, (2) view – displays the information to the user (more than one view may be defined), and (3) controller – handles the input from the user

Basing on major architectural patterns, students should provide an overall architectural diagram of their product, for instance, as the one shown in Figure 4.5.

Architectural view. Understanding the difference from design and its previous activity (analysis) and post activity (programming). Design is all about getting a vague system description, to a specific and detailed description that can be implemented and realized. A system can be viewed from multiple perspectives. One way to represent design is the 4+1 view model (Kruchten 1995). UML is often the notation used to visualize these views:

- **Logical view**: The logical view is concerned with the functionality that the system provides to end users. UML[1] diagrams are used to represent the logical view, including class diagrams and state diagrams.

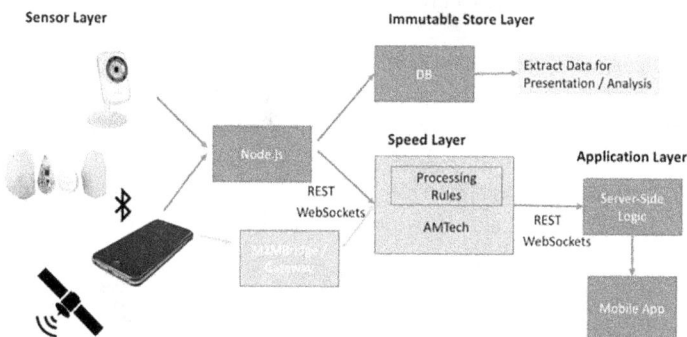

FIGURE 4.5 An example of an architectural overview of an Internet-of-thing application.

[1] https://en.wikipedia.org/wiki/Unified_Modeling_Language

- **Process view**: The process view deals with the dynamic aspects of the system, explains the system processes and how they communicate, and focuses on the run time behavior of the system. The process view addresses concurrency, distribution, integrator, performance, and scalability, etc. UML diagrams to represent process view include sequence diagrams and activity diagrams.
- **Development view**: The development view illustrates a system from a developer's perspective and is concerned with software management. This view is also known as the implementation view. UML diagrams used to represent the development view include package diagrams.
- **Physical view**: The physical view depicts the system from a system engineer's point of view. It is concerned with the topology of software components on the physical layer as well as the physical connections between these components. This view is also known as the deployment view. UML diagrams used to represent the physical view include deployment diagrams.
- **Scenarios**: The fifth view displays a small set of use cases with sequences of interactions between objects and between processes. They are used to identify architectural elements and to illustrate and validate the architecture design. They also serve as a starting point for tests of an architecture prototype. This view is also known as the **use-case view**.

Architectural tactics are different methods for achieving desired quality attributes. One can think of tactics as a catalogue of solutions to different problems addressing quality attributes. Many different tactics working together can often describe an architectural pattern. An Architecture tactic can be viewed as: stimulus→tactical design decision→estimated response.

In a modern applications, it is quite scarce that one will write code from scratch. Software are normally built on top of existing frameworks, or using libraries, as shown in Figure 4.6a (web frameworks and libraries) and Figure 4.6b (other frameworks, libraries, and tools). Therefore, the description of architectural patterns and views often bases on the architecture of these frameworks and libraries.

4.4.3 Implementation

General knowledge about programming is expected to be covered in previous courses. Depending on the actual project commitment, this project might require that the team members learn new programming language(s), new concepts of programming, various technical skills, etc. The group has to plan

jQuery	48.7%		Node.js	49.9%
React.js	31.3%		.NET	37.4%
Angular/Angular.js	30.7%		.NET Core	23.7%
ASP.NET	26.3%		Pandas	12.7%
Express	19.7%		Unity 3D	11.3%
Spring	16.2%		React Native	10.5%
Vue.js	15.2%		TensorFlow	10.3%
Django	13.0%		Ansible	9.4%
Flask	12.1%		Cordova	7.1%
Laravel	10.5%		Xamarin	6.5%
Ruby on Rails	8.2%		Apache Spark	5.8%
Drupal	3.5%		Hadoop	4.9%
			Unreal Engine	3.5%
			Flutter	3.4%
			Torch/PyTorch	3.3%

(a) (b)

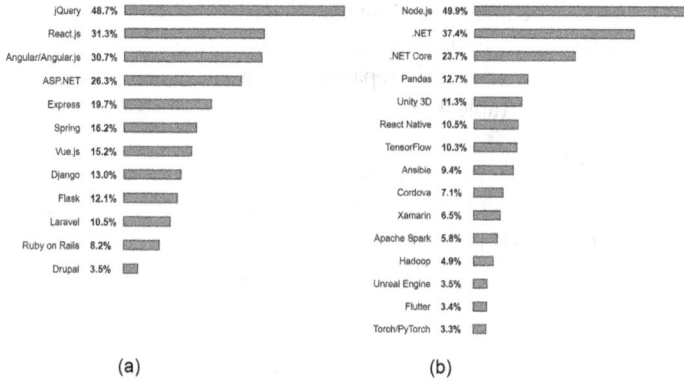

FIGURE 4.6 (a) Most popular web frameworks; (b) Most popular application frameworks.

how to obtain this knowledge, maybe in cooperation with the customer and the advisor. There are certain considerations before you start writing code. The code you write might be used as a base for further development and may also be used by the other team members.

- Inside the group, it will also be practical to have common design and code conventions that all group members understand and practice. If the customer has a coding convention, they probably would like you to use it also.
- Please observe that some freeware or trial-ware licenses of code editors, etc., state that it is prohibited to use them to write code for commercial use. Check the license of the software that you decide to use in the project and discuss it with the customer if there are such clauses that you might be in conflict with.
- Versioning control and backup tools, such as Git, should be used for all technical artefacts (UML diagrams, code, test data, etc.) and documents in the project. When you set up a development environment, make sure to set up a functional versioning system with backup as well.

4.4.4 Testing

Besides unit testing, other types of testing should be reported in this section. The testing, or quality assurance, is usually planned and carried out in four activities: (1) creation of a test plan, (2) specification of test cases, (3) execution of testing, and (4) report and approval of test results.

Test plan is a document detailing the objectives, resources, and processes for a specific test for a product. The test plan tells what type of tests are conducted, how usability tests are prepared. When you create a test plan, it is important to specify (Table 4.1):

- Which kind of tests should be carried out?
- What are the specifications of test cases?
- Who are the test persons?
- When should the tests be carried out?

TABLE 4.1 Types of tests

TESTS CARRIED OUT BY DEVELOPERS/TESTERS

Unit test (programming phase)	Testing of the smallest units in the projects, i.e., user interface, methods, stored procedures, objects, classes, etc.
Model test (design phase)	Entities integrated into bigger software components. Modules are tested to assure that the coordination and communication between the entities are as expected.
System test (requirement phase)	All modules that together form a complete version of the system should be tested. The system is tested to assure that the coordination and communication between models are as expected.
Integration test (design phase)	This is a complete test of the system and its interfaces to the world around. The last defects should be found, and it should be verified that the system behaves well according to the requirement specifications. In some projects, integration and system tests are merged.

TESTS CARRIED OUT WITH END USERS/CUSTOMERS

Usability tests (nonfunctional)	These are tests that assure that the interaction between users and the system is as expected. The goal is to get user-friendly applications.
Acceptance tests (nonfunctional)	Here, the end users should test if the system and its user interface to its environment are as expected. Based on this acceptance test, the management or customer makes decision on whether the product should be used or not.

Test cases are a specification of the inputs, execution conditions, testing procedure, and expected results that define a single test to be executed to achieve a particular software testing objective. The test case should cover all the user stories in the final product backlog. Test report summarizes the results of testing activities and to provide evaluations and recommendations.

Project closing

5

This chapter covers practical matters to complete the projects. It provides a specific instruction for project delivery (Section 5.2.1), project presentation (Section 5.2.2), and team reflection (Section 5.2.3) in the context of our project courses. Section 5.2 recommends a structure of the project report.

5.1 PROJECT CLOSING

A project or a project phase, after achieving its objectives or being terminated, will always need closure. Formal administrative closure consists of documenting project results to formalize acceptance of the project outcome by customers. It also includes collection of project records, analyzing project result, lessons learned, and archiving this knowledge for future use (PMBOK 2013). Project closures include major activities: (1) evaluate if the project delivered the expected benefits to all stakeholders, (2) assess what was done wrong and what contributed to successes, and (3) identify changes to improve the delivery of future projects. Within the course, project closure requires customer involvement and feedback, supervisor's feedback, delivery of both project report and source code, and final presentation.

One common mistake that many students face with in their early experience with project courses is at project delivery. One might think that when the final code is pushed and tested, he/she is done and only few things left so that they can switch to other courses. But is it ensured that the delivered code, products, documents, etc., will be well perceived by customers and lecturers? Furthermore, feedback is the central element of learning that needs to be established at both individual and team levels.

5.1.1 Project delivery

We have experienced from year to year many students losing points due to improper deliveries. While grading criteria do not explicitly mention about deliveries, any course instructions that are not followed imply reduction of student's grade. Common mistakes include (1) missing required artifacts, missing group number and group member names in file names or inside documents, (2) wrong file formats. For instance, at our course in 2019, we require final packages to include:

1. Final report as pdf files. Format: GroupNumber_Report_PRO1000. pdf (example: Group03_Report_PRO1000.pdf)
2. Final presentation slide. Format: GroupNumber_Slide_PRO1000. ppt (example: Group11_Slide_PRO1000.ppt)
3. Link to the demo video. Format: GroupNumber_Video_PRO1000. txt (example: Group07_Video_PRO1000.ppt)
4. Source code, installment instruction, and any relevant materials. Format: GroupNumber_Code_PRO1000.rar (example: Group08_ Code_PRO1000.rar)

These four materials should be put in one rar file. Format: GroupNumber_ PRO1000_2020.rar (example: Group01_PRO1000_2020.rar)

5.1.2 Final presentation

In our courses, there is a compulsory presentation for every team. Students are expected to perform the presentation together in front of public. Few pages of slides are also expected. They are often instructed about how they should perform the presentation in advance. Slides typically include information about:

- Team: photos and names of team members
- Agenda: what will be presented
- Project: brief introduction about the project and product
- Project requirements/Product backlog: highlight which items are finished, which ones remain
- Planning perspective: how the project was planned in terms of team, role, technologies, communication, and process
- Process planning: describe the team process, i.e., teamwork, agile practices
- Lesson learned: what are interesting lessons from the project?
- Product demonstration

5.1.3 Team reflection

Reflections can be done at both individual and team levels. At individual levels, each team member writes what they have learned from the project experience and what needs to improve for themselves and for the course. A team can discuss these points in among them and create a collective reflection that covers opinions from everybody. In a detailed manner, reflection can be written for the course, the project, the customer, the team, or the supervisor. Below we present some examples for each aspects, including:

REFLECTION ABOUT THE COURSE

As this project is the biggest and most extensive project any one on the team has been a part of, everyone in the team agreed that it was the most stressful, challenging, and time consuming project the team has been a part of. With that said, it was extremely educational and worth the time spent on it. The magnitude of what has been learned involving project management, team-work, programming, customer relationship, and development in general is a lot bigger than members of the group would have learned in the same amount of other courses.

(Group1_2017)

REFLECTION ABOUT THE PROJECT

In retrospective, we consider our project to be very interesting and fun to conduct. The applications that we are making will be used by actual end-users when our customer test it on their visitors. If the customer decides to continue using the application after the test phase summer 2016, then it could be used by lots of people.

(Group6_ 2015)

... and This project assignment was a good way of learning how to properly work on a bigger project as a team. The project was both stressful, challenging and fun at times, where in the beginning we didn't really know where to begin, or how important structure was for a project, until the end where we got a much better understanding on how we should work and structuring the project, from trial and error and coaching from the teacher.

(Group4_ 2020)

REFLECTION ABOUT THE CUSTOMERS

A central and important part of this project was the customer. The team had a very good communication with the customer from the very beginning. Except from one meeting, every meeting has been done over Skype, and other communication on Slack or email. The customer has been active and quick to respond on every issue or question from the team

(Group2_ 2013)

REFLECTION ABOUT THE TEAM

Certain members have had a difficult time choosing their own tasks for the project, or not chosen anything at all if it was not given. Some also need a lot of guidance to get started with what they were working on sometimes. This is probably because we have let everyone choose what they thought would be a good fit for themselves to do. Next year we will have a clearer leader role which will give everyone a Task for each sprint, make sure everyone gets started and always have something to do.

(Group2_ 2020)

AND FINALLY REFLECTION ON THE SUPERVISOR

It is clear that the project would have been difficult to complete with the same level of quality without the advisor, simply due to complexity of the project and course itself. The advisor was an indispensable asset to the team, filling her role as a one-person "steering committee" admirably. The helpful feedback which was provided was deemed highly valuable as it helped optimization of the process and gave a new perspective in different cases.

(Group1_2013)

5.2 REPORT GUIDELINE

This section recommends a comprehensive structure for reporting a software projects for CS&E students. We expect third-year students and above be able to follow this guideline. For reports from freshmen, the core parts of the report, from Section 5.2.4 to 5.2.12, should be covered.

5.2.1 Cover page

The cover page, also called as a title page, is the first and front page of the report. It is an important part of the document as it gives the introductory information regarding what the document is about as well as who has written it. A cover page for the project course should include the following information:

- Project title
- Course name
- Group number
- Full name of group members
- Submission date
- Logos of the University and Customers (if available)

The design of the cover page should be clear but also attractive to readers. Consider using block of texts, different font styles, page layout, colors, and logos.

5.2.2 Abstract

Abstract is the next page of the document, after the cover page. An abstract is a short and concise summary of the report, containing all the important highlights of the document. The length of a report abstract should range from 200 to 350 words. We recommend students to use a structured abstract for project report, as below:

- [Context]: three to five sentences stating the importance of the problem addressing in this project
- [Objective]: one or two sentences stating the scope of this project
- [Execution]: three to five sentences stating how the solution is implemented, i.e., what are technology stacks, what are the software development approaches, how the teamwork is performed
- [Results]: two to five sentences summarizing what are achieved by the end of the project, feedback from customers, supervisors, and future perspective
- [Acknowledgement]: this is an optional part in case there needs to thanks to external people who help to carry out the project

The content of the abstract should be put in a separate page with explicit title "Abstract."

5.2.3 Indexing pages

Indexing pages include Table of Content, List of Figures, List of Tables, and Terminology. Only Table of Content is required, other lists are options, depending on how many items are there in the report.

Table of Content page should come right after the Abstract. The Table of Content can be easily generated by modern word processing software, i.e., Microsoft Word or Latex. Some tips when creating the Table of Content:

- Choose the style that is consistent with the whole report, in terms of font size, layout, etc.
- Table of Content should not be longer than four pages.

5.2.4 Introduction

The introduction tells the readers what the report is about. It sets the project in its wider context and provides the background information the reader needs to understand the report. We recommend students to include the following content, typically as subsections of the Introduction as below:

- [Project background]: this is often an extended version of the project description given by the customer. The student team investigates more by searching themselves or talking to the customers and describes the organizational, social, and technical contexts of the project.
- [Project problem]: description of the problem will be addressed in the project in a separate paragraph. Also state the goals or objectives of the project and the expectation of the customers.
- [Project stakeholders]: presentation of key people involving in the projects
  - The group: each paragraph presenting information for each member of the group. Contact information should be included.
  - The supervisor: name and contact information.
  - The customer: the organization of the customer (as specific as possible), the type of the organization, what product/service they provide, the name and contact information of the customer representative.

- [Scope and constraints]: a paragraph saying how long the project will last, what are the known and unknown constraints the team needs to consider.
- [Work distribution]: this is an important section. The team should be able to report how much each team member contributes to the project, in terms of types of tasks and working hours.

5.2.5 Project planning

The section should reflect the last version of their project plan. We recommend students to report the following aspects of the plan, typically as subsections:

- [Scope planning]: defines boundaries of the project, what will be done and what will not be done in the project, and what will be delivered and what will be not by the end of the project. Work Breakdown Structure (WBS) is the essential artifact of this subsection.
- [Time planning]: scheduling and setting priorities to activities in the projects. Important milestones should be described. Gantt Chart is the essential artifact of this subsection. Activities in Gantt Chart are often taken from WBS.
- [Project organization and communication]: what are the roles defined in the project? Who will take which role? What kind of meetings are there in the project? How will they be conducted? What are communication tools? How does the team plan to use them? What are other collaborative tools, such as text editors, version control systems, etc.?
- [Risk planning]: students report the list of risks associated with the project, their frequency, severity, risk value, and countermeasures. A risk management table should cover all of the previous points in a table format.

5.2.6 Prestudy of problem space vs. solution space

The prestudy phase, or in some places called Sprint 0, is the period of exploration; the students should formally plan for their exploration of the problem and the possible solutions. The students need to discuss within their team what need to be learned and who will do what. The structure of this section can be organized as below:

- [Limitation]: what is the limitation in terms of time, knowledge, skills, infrastructure, regulations, etc., at the beginning of the project? How does the team plan to address these limitations?
- [Learning plan]: what knowledge/skills should be acquired during the project? How will the learning be performed?
- [Possible solutions]: what are the possible solutions in terms of technical and organizational aspects? How does the team plan to try out?

5.2.7 Software development methods

A software development methodology is a framework for developers to be able to plan, structure, and control the different phases of the process of software development. Students should be able to reason about what development approaches they will use and provide the description of the actual approaches they used throughout the project. The presentation of this section should include:

- [Overall methodological description]: for instance, if Water-fall, Agile, Test-driven development, or Lean startup, etc., will be used
- [Rationale for the choice]: the students should be able to justify why the methodology is suitable to the project, given the project requirements, constraints, and the team capacity.
- [Detail description of practices and techniques]: the students should describe what and how they use software development processes, practices, and techniques. For instance, a description of Retrospective Meeting and how the meeting is organized should be given.

5.2.8 Product backlog

Students are encouraged to specify the product requirements as user stories. User stories are short feature descriptions taken from a user's perspective. The user stories were created as a result of discussions with the customer and are based on user scenarios. Note that user stories are different from use cases. A final product backlog that covers all initial requirements should be put in a table format as below (Table 5.1):

TABLE 5.1 Product backlog table template

BACKLOG ID	DESCRIPTION	ESTIMATED POINTS
Item1 (functional or nonfunctional item)	As a …(who wants to accomplish something) I want to … (what they want to accomplish) so that … (why they want to accomplish that thing)	(use working hours, planning poker points, story points …)

5.2.9 Architectural design

The section presents architectural and technological aspects of the product. Hence, we suggest having separate subsections:

- [Technology stacks]: includes programming language for front-end, back-end, database technology, cloud services, third-party APIs, open-source library and frameworks, etc. It is important to justify why the technology stacks are chosen.
- [Overall architectural patterns]: at first, a general architectural view should be provided. In many cases, the architecture reflects the adopted frameworks or libraries. Typical general architectures are client-server, n-tier architecture, micro-service architecture, model-view-controller pattern, etc.
- [Architectural views]: 4+1 architectural view model provides a standard way to describe multiple views of a software system. In case the students are not familiar with these views, it should allow a simpler version of the detailed architecture. At least three different views should be considered whenever the views are applicable to the selected solution:
  - Class diagram: a static structure of the product from an object-oriented perspective
  - Database diagram: a basic structure of data that is used in the product, including data tables or data objects and their relationships.
  - User experience design: a final screenshot of the font-end design for the product

5.2.10 Testing

Besides unit testing, other types of testing should be reported in this section. The testing section should include:

- [Test plan]: a simplified version of a test plan should be given at first. The test plan tells what type of tests are conducted, how usability tests are prepared.
- [Test cases]: a list of test cases with a representative test for each case should be provided. The test case should cover all the user stories in the final product backlog.

TABLE 5.2   Test case table template

TESTCASE ID	DESCRIPTION	ASSOCIATED USER STORY	RESULTS
Item1	Brief summary of what functional/nonfunctional features will be tested Detailed description of the test can be provided in a separate table if needed	Link to a user story Id from Section 9.8	Summary of the test result and comments

- [Test result]: the overview of the test results for each type of test cases should be given. Regarding usability test, qualitative analysis (summary of opinions from users via interviews) or quantitative analysis (aggregation of rates from users via surveys) of the results is expected (Table 5.2).

5.2.11 Sprint summaries

This section should be split into several subsections, each of them is for one Sprint, for instance, Sprint 1, Sprint 2, Sprint 3, etc. A brief and concise description of each Sprint covers information as below:

- Sprint goals: what is the main purpose of the Sprint? (to learn the solution space, to set up the development environment, to develop the key user stories IDxxx, to focus on usability testing, etc.)
- Sprint backlog items: A similar table with Table 2.1 but at Sprint level.
- Sprint review meeting: one paragraph summarizing which tasks are finished and which tasks are not in the Sprint
- Sprint retrospective meeting: one paragraph summarizing what went well, what went bad, and what to try in the next Sprint.

5.2.12 Project evaluation

This section describes the group evaluation and reflection of the working process throughout the whole project. The purpose of reflection is for the future improvement from both students' and teachers' sides. We recommend having two separate reflections:

- Reflection on the students' side: teamwork, process, and project management
- Reflection on the teachers' side: course arrangement, customers, projects, lectures, and supervisions.

5.2.13 References

It is important to properly and appropriately cite references in scientific/ technical documents (such as books, websites, blogs, papers) in order to acknowledge your sources and give credit where credit is due. Examples of how citations and references should be used can be seen from this Guideline. There are many different formats for your references. You need to be consistent with one formatting style. See more here: http://tim.thorpeallen.net/ Courses/Reference/Citations.html

We recommend students to use one of the reference management software: Zotero[1] or EndNote.[2] The software is easy to use and offers plug-ins for Microsoft Words.

5.2.14 Report format

There is not strict requirement on the length of the report. In case of page length limitation, students can provide the content as external materials and provide links to the materials. The report should be submitted in the **PDF** format, following the instruction from the Exam office. The report should not contain:

- Large images that contain unreadable texts. All images inserted in the report need to fit to its layout (either in horizontal or in vertical way).
- Images that copy and paste directly from i.e., Trello, Kanban board, or screenshots of other software tools. Images should be exported properly from those tools or redrawn to fit to the report.

---

[1] http://zotero.org/
[2] https://endnote.com/

Incremental project-based learning

6

6.1 MOTIVATIONS TO THE INCREMENTAL PROJECT-BASED LEARNING

The project-based learning in computer science and engineering discipline is widely used technique to teach technical subjects. The objective is to help students implement the theoretical learning into meaningful projects in academic settings. The projects associated with the individual courses motivate students to take their learning into suitable level in order to successfully handle complex projects. The individual course project experiences help a student to gain practical exposure and unify his learnings toward completion of capstone project in his final year. The capstone project is a complex project that tests the throughout degree learning of the student and requires multiple people with multiple roles and specializations.

The student could implement database project, programming language project, data analytics project, software engineering project, or so on. Each project is based on the project idea i.e., the solution that solves the given problem. Approaches for handling such projects differ; for example, software engineering models such as Agile model are used to handle software development projects. The ideas for such projects could come from industry, faculty, personal observations, research papers, etc. However, developing a software solution based on the identified project idea is never a "one go" solution. In fact, the solution is delivered in increments with each increment implementing new functionality along with improvements. The incremental approach helps software team to handle dynamism in markets and uncertainties about the software (for instance, uncertainties about the problem domain, solution domains, and markets).

Another observation in computer science and engineering programs is the existence of similar project-based courses at different academic years. Some courses have almost the same format of a semester-long software development projects with some specialized topics, such as Agile Development, Security, and Software Engineering. Some courses are carried on at a faculty level with students from various programs. We have seen that the learning experience from these courses is often isolated from each other, and students ending up with a long list of projects they need to do throughout the whole academic program. With a possibility of coordinating different project-based courses in the same program, we can avoid boringness and repetition, adding deeper, additional learning experience to students. In a specific context, for example, self-defined projects, students even have enough time and space to explore and to develop some solutions, which is good enough to have practical influence in real life.

Incremental project-based learning helps to enhance the software engineering related knowledge of the students by helping them learn in real uncertain dynamic situations and implement the learning to improve later increments. For instance, continuously working with incremental development will help them understand the evolving nature of customer requirements, importance of documentations and the tools, etc. This makes them able to handle the industrial problems because of their experience with evolving software development, better exposure to the research domains, and better access to technology learned in academics. Finally, the learning could help them implement niche innovative capstone project that is either an innovative version of old project or a completely new project. The various research papers, technical reports, experience reports, etc., will help the student to share his learning with external world, get access to outside world feedbacks, and establish ownerships to the solutions.

Initially when the student team approaches the project idea, the level of uncertainties is quite high. This makes it hard to accurately predict the market of the software solution and accurately plan the release. Once the first increment is delivered (or submitted as technical reports in academics), the student gains enough level of understanding not only with the problem domain but also with the technical aspects, as a result of exposure to new technologies, identifying scope of improvements (for instance, applying new technologies or research solutions to the problems), and customer feedbacks. This software release after evaluation by the evaluation committee and use in real context provides rich feedback to the students, which helps them to evolve the software further (Increment 1). The learning gained helps the students to gain the better understanding of the market, lower uncertainties about the product,

er8ng

and enhance accuracy of planning, which leads to second incremental version of the software (Increment 2). Second increment is the better version of the previous increment with lot of performance improvements and new functionalities. The innovation of the software product continues with the enhanced learnings, leading to the new incremental versions of the software solution. The incremental versions are usually released by the student team one per each semester as they try to innovate existing solutions by including new learning brought as a result of their exposure to the software users, new skills gained through internships, or new courses.

Figure 6.1 captures the essence of the iterative project developments through a spiral model. The project development is conducted in number of increments (denoted by 1, 2, 3 and so on) with each individual incremental development composed of three stages i.e., project planning, project execution, and project closing. Each stage is represented by the quadrant in the spiral model. Each iteration delivers the product with functionality better than the previous increment. The suggested artefacts for each project development stage are highlighted in Figure 6.1 and discussed throughout the book. The type of artifacts remains the same for each iteration although each artifact becomes more and more complex with the number of iterations. This is because with each iteration, the customer experiences the new functionality and their needs always change. This requires software team to incorporate

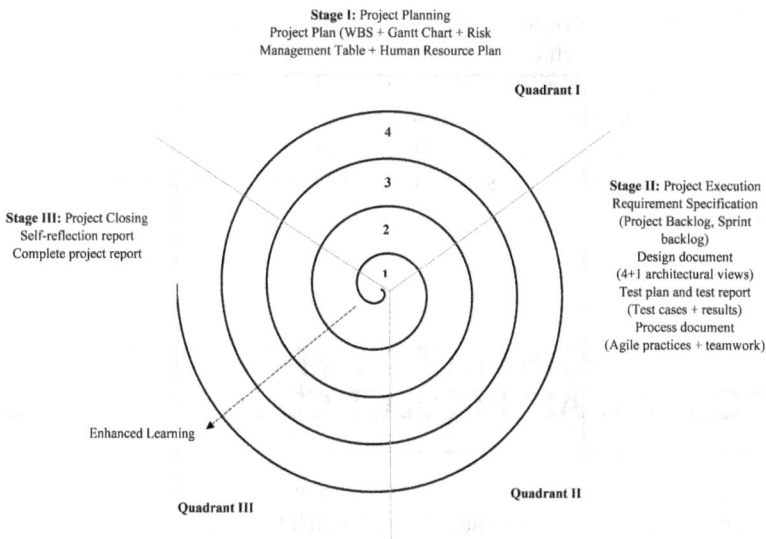

FIGURE 6.1 Multiple iterations of project-based learning courses.

their feedback that represents new functionality, change requests, or deletion requests. With each academic semester or year, the students could measure their project success in the market (or academic settings) to identify the innovative ways to increasing the value delivered by the software. This enhanced learning helps the team to keep increasing the value proposition of the software product. Finally, the outcome of each incremental development is the project report, and students could also decide the possible venues for publications of useful results (see Chapter 7 for details).

6.2 ITERATION 1 OR "CLASSICAL" PROJECT COURSE SETTING

First iteration is the first exposure of the students in the software development world. In this stage, the requirements are usually expressed by the project supervisor or are derived by their observation of the real-life scenarios. Overall, the experience of the students in managing the software development and their understanding of the problem domain is limited. For instance, their risk may not be accurately identified and ranked. They may not have strong contingency risk plans. They may take too much time in designing design documents, etc. However, they get access to the exposure to the simulation of the industrial software development. The software solution developed by them could invite rich feedback from the evaluation committee (also including users, if any). The exposure of the students to the advanced computer science courses will also help them generate creative ideas about improvements that could be made in the software solution. The team will have serious difficulties in creating the software artifacts, which may also be not 100% accurate in this increment.

6.3 ENHANCED ITERATION OR "CLASSICAL" PROJECT COURSE SETTING

In future increments, students gain experience with the software development, computer courses, software markets, and high technological knowledge. This helps them to continuously use their enhanced learning to evolve the software artifacts based on real market facts and continuously improve their software solution. This enhanced learning helps them to improve their project planning, execution, and closing activities as discussed below:

6.3.1 Project planning

The increased exposure to the continuous software development activity helps the team to improve their project planning activities. This is because, they get exposure to the market facts, which were mere assumptions before. These facts help them to increase the effectiveness of their planning by learning through their mistakes. The artifacts such as WBS, Gantt charts, communication plans are all based on the improved learning of the team members.

- Scope planning: The scope of the project always increases as there are a number of requests for new functionalities, change requests, and improvements to be made. With the delivery of new increments, the complexity of the software increases.
- Time planning: The WBS and Gantt chart need to be created afresh, but they will be benefitted by the reuse of experiences with previous iterations in identifying tasks and their durations, etc. The experience with the previous incremental developments could help the team to better estimate the work to be accomplished (and their subtasks), their durations, and efforts (usually in person-months), which could help them prepare accurately the WBSs and the Gantt charts.
- Risk planning: The risk estimation is purely dependent on the expertise of the project manager. In academic settings, the project supervisor helps the students to plan for resolving few risks that could become reality in the future. However, risk identification and their ranking are purely based on the estimation of the project supervisor based on his previous experience. As the team gains learning with incremental development, their predication about the risks becomes more and more accurate. The team gains expertise in accurately planning for a wide array of the risks and effectively executing/modifying the resolution plans if risks become reality.
- Resource planning: The incremental development helps the team to identify the skill gaps in the team for which they could plan trainings. The updated skills help to improve the software development outcomes in terms of improved development times, costs, and quality.

- Way of working: With each increment, the team always learns through their mistakes. They could identify the better model to execute software development activity or could optimize existing model as per their needs.

6.3.2 Project execution

- Redefine the problem space and solution space: With the new learning, the understanding about the problem domain increases. The team knows better the problems faced by the user. Users also gain experience working with the software solution that helps them to better express their unmet needs to the requirement engineering teams. The software solution gains better acceptability among the users because of the better understanding about the problem domain. The solutions designed to address the user problems may not work perfectly well. With each increment, the team could try to optimize the solution and improve its performance to make it easily accessible to the users.
- Process monitoring: The improved learning helps software team to gain experience about the best ways to record their hours and planning and conducting meetings with the team members. This helps them to establish the better coordination mechanism based on the mistakes they did in the previous increments.
- Software engineering: The software team usually uses the same software engineering process model as used in previous incremental developments. This helps them to gain expertise in execution of the same model and handle uncertainties in effective manner. However, the optimization of the software engineering process models could also be possible by incorporating the best practices and adopting them to their working context.

6.3.3 Project closing

- Quantify the gained learning: The students with each increment gain exposure to practical software development in industries and their evolution driven by the customer feedback. They also learn that to remain competitive, they must maintain innovation in their product. Also, the market is always dynamic, and changes are the realities of the actual developments.

- Quantify the gained value for customers: Each increment gives the user the better experience in terms of improved performance and higher functionality. The continuous innovations are always adding value to the users.

- Reflecting on new learning from students: Students should be able to reflect on not only what they learn via multiple iterations in this course but also experience from previous project instances, i.e., in a long-term perspective, what they have improved and what stay as "persistent" issues?

- Reporting: Team could use same report format as used previously. But, report of the new increment should explicitly refer to the previous report to create virtual links between them, thereby promoting traceability between the knowledge shared by each report.

Dissemination of results

7

The dissemination of the work conducted by the students in various venues such as Fair, Industrial events, Journals, and Conferences helps them to share their work with the audience and get access to the high-quality feedback that will lead to product innovation. This also helps the students to establish the ownership over the work that they are sharing and establish the credibility of the work. The dissemination is purely optional but highly recommended by the universities these days. This chapter has two objectives: (1) providing guidelines for selection of the possible publication venues and (2) providing guidelines for deciding suitable type of research articles and using suitable article format.

7.1 INTRODUCTION

The dissemination of the project outcomes such as experiences, innovative method used (for instance, research solution to the problem in hand), validated solutions (research solutions validated in laboratory settings), working code, etc., in various venues such as Conferences, Journals, and Patents, helps the students to share their research with the other researchers. Not all the projects could be research-based. For instance, students could choose to target a known problem with the known solution available in the literature. This means that the solution to the problem is well known and the students implement them using suitable computer language. Such projects with little research are usually during initial years of the bachelor's degree, which aim to provide students with the exposure to the real projects. The potential candidates for dissemination are usually experiences and the working code.

For research-based project, the students identified an innovative solution for the problem in hand, which is finally implemented using suitable computer language. Disseminating the outcomes of the projects helps your research reach the target research audience and helps by providing the future direction to the research with valuable feedbacks. The ability of the students to publish their research in leading venues establishes the credibility, accuracy, and quality of the research work. The publication also helps them to establish contacts with industries, research units, and higher education institutions for internships, jobs, and education. Registering the research-based working system as patents not only helps students to establish ownership of the work but also commercialize it. There are numerous venues for publications and different types of research articles that could be published.

The dissemination of the aspects related to the project helps the team to gain access to the feedback of the external world, which is advantageous to innovate project, share the project-related aspects with external world, and get ownership of the project completed. The decision to whether the project-related aspects to be disseminated or not depends on various factors such as permission of the sponsor (if the project is sponsored by sponsoring agency such as industry or NGO, etc.), university policies (usually publication is very supported by university), faculty guide permission, etc. Also, the names of the authors to be included in the research publications and the order of their appearance must satisfy the ethical practices in research. For instance, the involvement of an expert at advisor level does not entitle him to be an author of the publication, but he can find his place in the acknowledgment section.

7.2 PUBLICATION VENUES

There are different venues for publication of the research work. These days there are plenty of such venues, and selecting the best one strongly depends on the quality of the work carried out, type of work to be disseminated, and ownership issues. This section provides details of possible venues and the criteria to be used for selection of such venues for the dissemination of the project work. The software projects being global in nature must motivate the research team to disseminate project findings with international community of researchers. Thus, only venues at international level are considered in this book.

   a. Conferences
      International conferences are organized throughout the world, and proceedings are published by leading publishers such as IEEE, ACM, Elsevier, and many more. Few conferences publish the

accepted and presented papers as conference proceedings directly available in bibliographic databases (for instance in IEEE Xplore), some publish them as edited book and then make the entire book available in bibliographic databases (for instance, Lecture Notes in Computer Science published by Springer and made available on SpringerLink), as journal article (for instance, conference proceedings in Procedia Computer Science Series, Elsevier), etc. Besides this, the authors are also invited to submit their papers in leading journals as extended versions of their conference papers. The authors benefit from fast review process of the extended special issues. The conference helps the researchers to interact with other researchers and share knowledge with each other. Due to the advancement of Information and Communication Technologies (ICT), the researchers are also allowed to present their papers through online tools such as Skype, Microsoft Teams, or Google hangouts.

International conferences could be ranked differently by different evaluators, for instance, Clarivate indexes few conferences using Conference Proceedings Citation index,[1] Computing Research and Education Association of Australasia (CORE)[2] ranks conferences using A*, A, B, C, and other ranks. Higher the rank, higher is the quality of the conference.

There are numerous parameters that could be used for selection of the conference such as its location, registration fees, ability to physically attend conference, etc. However, the prominent factors recommended include the following:

- Reputation of the publisher: For instance, proceedings published by leading publishers such as Taylor & Francis,[3] IEEE,[4] Springer,[5] and ACM[6] signify high quality of conference and global coverage of the proceedings.
- Indexing of the proceedings: The proceedings published and indexed in Scopus[7] are considered as a necessary parameter these days in academics and research institutions.
- Scope of the conference: Generally, the topic and article type must be within the scope of the conference.

---

[1] https://clarivate.com/webofsciencegroup/solutions/webofscience-cpci/
[2] https://www.core.edu.au/
[3] https://taylorandfrancis.com/
[4] https://www.ieee.org/
[5] https://www.springer.com/
[6] https://www.acm.org/
[7] https://www.scopus.com

- Highly renowned conference committees such as International Advisory Committee, Reviewer Board, etc., imply higher review standards of the conference.
- Acceptance rate of the conference: The low acceptance rate of previous editions of the conference signifies higher quality of the conference.
- Tie-ups with high-quality journals also signifies higher quality and opportunity to submit extended version of the conference paper to the journal.

b. Journals

Journals provide good opportunity to disseminate the project work. There are a wide number of international journals in different subject areas, based on different interesting topics, published by different publishers and indexed in different indexing sites. Depending on the type of article, length of article, and quality, different journals could be selected. Prominent selection criteria include the following:

- Reputation of publisher
- Option for traditional publishing vs. open access (both are equally good)
- Indexing of the journal (Science Citation Indexing, Social Science Citation Indexing, Scopus, ABDC Indexing are considered good)
- Impact factor of the journal
- Acceptance rate of the journal
- Free publishing (if not open access).
- Review time (should not be very less, for instance, less than a week)

c. Copyrights and Patents

Intellectual property rights are protected by copyrights and patents. What could be patented or copyrighted depends on regulations of the country. However, they prevent the unauthorized use, copying, and selling of intellectual property without owner permission. For instance, in India, the copyrights establish the ownership of the owner over his work (such as books). This prevents others from using the copyrighted work without permission of the owner. Patents on the other hand protect the working novel systems. Under this definition, the computer program could only be copyrighted and not patented. Further fees associated with copyrights are very less as compared to patent filling and process complexity also differs considerably. So, if the project work is to be protected from preventing others from copying the work, then copyrights are the best option. However, ideas can never be protected (except their written text). They could be patented as provisional application (or complete application if entire working system is available).

7.3 ARTICLE TYPES AND FORMATS

There are different types of articles that research team could select to disseminate their project work. This includes the following (as per Wieringa et al. 2006):

- Solution papers
- Validation research papers
- Evaluation research papers
- Opinion papers
- Experience reports
- Philosophical papers.

Solution papers are the research papers with new solution for the problem with no strong validation or evaluation, a small hypothetical example could be a part of it. **Validation research** papers are the papers that disseminate the solutions that are validated using experimental data, simulations, or in laboratory settings (and not in real scenarios). Stated in other words, the solution is not implemented in practice. **Evaluation research papers** include those that disseminate the solutions that are validated in real scenarios. Stated in other words, the solutions are implemented in practice. This category also includes the empirical studies that report the investigation of the existing solutions in practice (for instance, case studies about requirement engineering in software company, survey of existing requirement engineering practices, etc.). **Opinion papers** reflect the opinion of the researcher, **Experience research paper** reports the personal experience of the researcher i.e., lessons learned, and **philosophical papers** present a new insight into existing thing such as new conceptual model.

Each of the research article types has a different format that presents the research in well-structured and organized way. This book recommends formats for different research types. For the sake of simplicity, we divide the structure of the paper into two sections, i.e.,

a. title section and
b. content section

Title section contains information about title of the paper, subtitle (optional), and author details such as names, affiliations, email ids, telephone numbers, details about corresponding author, etc. Content section contains the actual research content that researchers wish to disseminate to the research audience. For all the different types of research, title section remains the same but content section varies.

Structure of title section (common for all types of research) is given below:

- Title of the paper
- [Subtitle (optional)]
- List of author names (separated by comma) with numeral subscript that uniquely establishes relation with the affiliation.
- List of author affiliations with numeral subscript that uniquely establishes relation with the author names.
- Author email ids (separated by comma) with numeral subscript.
- [Contact information (optional) (separated by comma) with numeral subscript.]
- Statement denoting the corresponding author (usually preceded by number or symbol that matches with subscript following the author name).

Content section varies considerably with the type of research. The following text summarized the format of content section but that's only recommendation based on the author experience with publishing research articles with leading venues.

Content section for solution papers

- Abstract
- Introduction
- Literature review
- Proposed solution
- Hypothetical example/working prototype
- Emerging results
- Discussion (optional)
- Conclusion and future work
- References

The various elements of the research paper are briefly explained below:

- **Abstract:** The abstract is the brief overview of the research paper. This section should be able to explain entire paper in a concise way. It is a short summary of the entire paper that helps the reader get overview of the research disseminated in the paper. This section helps the reader to decide if the research paper is of importance to him. Abstracts could be written as a short summary highlighting overview of the research conducted, results, and conclusion. To better enhance its readability, use of structured abstracts is preferred. Structured abstract is structured as background, research objectives, research method, results, and conclusion. The details provided against each section are concise enough to give a brief overview but formulated carefully to avoid any misinterpretations.
- **Introduction:** This section provides brief overview of the research area, brief details about motivations for conducting research in the area by briefly highlighting research gaps and structure of the paper. Research gaps are described in a concise way as they are provided in detail in literature review section.
- **Literature review:** This section provides the evidences from the literature about the state of art in the research area. These evidences are arranged in the manner to highlight the gaps that need effective solutions, which forms the basis of the conducted research. The proposed solutions are validated to ensure that identified gaps are bridged effectively.
- **Proposed solution:** The solution proposed by the researcher is mentioned here. Usually the solution is formulated as an algorithm using different conventions such as pseudocode. Suitable designs such as circuit diagrams could also be used if research solutions are based on such circuits. Coding statements could also be mentioned along with the flow charts depending on the nature of the solution.
- **Hypothetical example:** Here researcher should mention the hypothetical example that is used to test the working algorithm. In computer science, the solutions could be small hypothetical test suits, which should be clearly described.
- **Emerging results:** The solution is not tested in real scenarios, nor is it tested using effective simulations, larger test suites, or through working prototypes. In fact, they are tested through small toy examples, which does not ensure their effectiveness. The results

are thus emerging and may require evolution of the solution. Emerging results should be mentioned here.

- **Discussion (optional):** Researchers could lead discussion here that what were the gaps to be bridged, how the solution bridges those, and what they predict about their scalability.

- **Conclusion and future work:** Here researchers should provide meaningful conclusion about their research and provide future work including the need for strong validations.

- **References:** The evidences mentioned throughout the paper should be referenced here using suitable format. There are multiple formats and one the researcher selects depends on the one allowed by publishing venues.

Content section for validation research
- Abstract
- Introduction
- Literature review
- Proposed solution
- Validation approach
- Results
- Discussion (optional)
- Conclusion and future work
- References

The various elements of the research paper are briefly explained below:

- **Abstract:** The abstract is the brief overview of the research paper. This section should be able to explain the entire paper in a concise way. It is a short summary of the entire paper that helps the reader get overview of the research disseminated in the paper. This section helps the reader to decide if the research paper is of importance to him. Abstracts could be written as a short summary highlighting overview of the research conducted, results, and conclusion. To better enhance its readability, use of structured abstracts is preferred. Structured abstract is structured as background, research objectives, research method, results, and conclusion. The details provided against each section are concise enough to give a brief overview but formulated carefully to avoid any misinterpretations.

- **Introduction:** This section provides brief overview of the research area, brief details about motivations for conducting research in the area by briefly highlighting research gaps and structure of the paper. Research gaps are described in a concise way as they are provided in detail in literature review section.

- **Literature review:** This section provides the evidences from the literature about the state of art in the research area. These evidences are arranged in the manner to highlight the gaps that need effective solutions, which forms the basis of the conducted research. The proposed solutions are validated to ensure that identified gaps are bridged effectively.
- **Proposed solution:** The solution proposed by the researcher is mentioned here. Usually the solution is formulated as an algorithm using different conventions such as pseudocode. Suitable designs such as circuit diagrams could also be used if research solutions are based on such circuits. Coding statements could also be mentioned along with the flow charts depending on the nature of the solution.
- **Validation approach:** Here researcher should mention the details about the technique used for the validation of the research solutions. This could include details about simulations, working prototypes, and larger test suites (and underlying testing environment) used to test the solution.
- **Results:** The results of validation should be mentioned here. It is recommended to test solution using data sets of different complexities and sizes to ensure results replication. The results should provide evidence that solution is effective enough for different set of inputs. It is also recommended to provide link to working solution and data set to allow research community to replicate the results.
- **Discussion (optional):** Researchers could lead discussion here that what were the gaps to be bridged, how the solution bridges those, and what they predict about their scalability.
- **Conclusion and future work:** Here researchers should provide meaningful conclusion about their research, its applicability to solve real problems, and directions for the future work including the need for validations in real contexts (for instance, in industrial settings).
- **References:** The evidences mentioned throughout the paper should be referenced here using suitable format. There are multiple formats and one the researcher selects depends on the one allowed by publishing venues.

Content section for evaluation research papers
- Abstract
- Introduction

- Literature review
- Proposed solution
- Validation approach
- Results
- Discussion (optional)
- Conclusion and future work
- References

The various elements of the research paper are briefly explained below:

- **Abstract:** The abstract is the brief overview of the research paper. This section should be able to explain the entire paper in a concise way. It is a short summary of the entire paper that helps the reader get overview of the research disseminated in the paper. This section helps the reader to decide if the research paper is of importance to him. Abstracts could be written as a short summary highlighting overview of the research conducted, results, and conclusion. To better enhance its readability, use of structured abstracts is preferred. Structured abstract is structured as background, research objectives, research method, results, and conclusion. The details provided against each section are concise enough to give a brief overview but formulated carefully to avoid any misinterpretations.
- **Introduction:** This section provides brief overview of the research area, brief details about motivations for conducting research in the area by briefly highlighting research gaps and structure of the paper. Research gaps are described in a concise way as they are provided in detail in literature review section.
- **Literature review:** This section provides the evidences from the literature about the state of art in the research area. These evidences are arranged in the manner to highlight the gaps that need effective solutions, which forms the basis of the conducted research. The proposed solutions are validated to ensure that identified gaps are bridged effectively.
- **Proposed solution:** The solution proposed by the researcher is mentioned here. Usually the solution is formulated as an algorithm using different conventions such as pseudocode. Suitable designs such as circuit diagrams could also be used if research solutions are based on such circuits. Coding statements could also be mentioned along with the flow charts depending on the nature of the solution.

• **Validation approach:** Here, the researcher should mention the details about the technique used for the validation of the research solutions in real context. This could include details about real environment where solution is put for actual use and which entities were responsible for measuring its effectiveness.

• **Results:** The results of real validation should be mentioned here. It is recommended to highlight the actual results obtained with the deficiencies, if any.

• **Discussion (optional):** Researchers could lead discussion here regarding what gaps need to be bridged, how the solution bridges them, and what they predict about their scalability. The outcome of validations should also be discussed; i.e., whether it is above expectations or below. Reasons for the observations and scope for further improvements should also be mentioned here.

• **Conclusion and future work:** Here researchers should provide meaningful conclusion about their research, its applicability to solve real problems, and directions for the future work including the need for further evolution of the solution.

• **References:** The evidences mentioned throughout the paper should be referenced here using suitable format. There are multiple formats and one the researcher selects depends on the one allowed by publishing venues.

The opinion, experience, and philosophical papers are usually short-length articles written by experienced researchers. The objective is to provide to the research community the very deep expertise in the very important issue that is meaningful to the research audience. For the students, such papers could be submitted to the venues that especially ask students to share their opinions, experiences, and new conceptual model findings. Structure of such papers is highly motivated by the experience of the researcher, topic under investigation, and venue for the publication.

Content section for opinion papers
• Introduction to the topic under investigation
• Discussion about relevant research already conducted
• Discussion about innovative ideas that extends the conducted research
• Conclusion
• References

Content section for experience research paper

Experience reports help researchers share their experience in conducting research or their industrial experience that will benefit other researchers. Hence the focus of such reports is more on the experience to be shared and setting directions that how it will benefit the other researchers.

* Abstract
* Introduction
* Background
* Description of the problem and related context
* Formulation of experience-rationale, working context, solutions, lessons learned, tools, techniques, practices, etc.
* Conclusion
* References

Content section for philosophical papers

* Introduction (mention here the arguments that you are going to defend, oppose with claims)
* Background
* Discussion on claims

7.4 ARTICLE TYPE, SUGGESTED VENUES ACADEMIC YEAR WISE

Publishing in leading venues is an optional activity. The students are expected to submit their project reports after the completion of their projects. However, to foster the innovation in the projects by inviting external feedback and establish the ownership of the project work, it is advised to undertake the dissemination tasks as one of the primary activities during project undertakings.

S. NO.	ACADEMIC YEAR	ARTICLE TYPE	ARTICLE LENGTH	SUGGESTED VENUE	MANDATORY
1.	Year 1	Solution paper Opinion paper	Short	Conference (emerging result/vision/ short communication)	Technical report
2.	Year 2	Solution paper	Medium	Conference (tool/research track/emerging results)	Technical report
3.	Year 3	Validation paper	Medium	Conference (tool/research track) Or Journal (research article)	Technical report
4.	Year 4	Evaluation paper	Long	Journal (research article)	Technical report
		Experience paper	Medium	Conference Or Journal (short article)	Technical report

Appendix 1
Project description

PROJECT 1 – FOODSNAP MOBILE APPLICATION

In modern societies, the majority perceive healthy eating as complicated. Seeking comprehensible and actionable advice, Americans spend over $40 billion each year on diets and self-help books but achieve little success: the majority eventually regain any lost weight and more. For some diseases, it is critical to control the amount of intuition taken in daily basis.

Recent advance in Artificial Intelligence makes highly accurate recognition of food via photos. Automated estimation of nutritional amount via daily food has been feasible. In this project, you will develop a web app that is able to recognize photos of food. The photos can capture fruits (apple, orange, etc), common dishes, i.e., pizza, spaghetti, taco, etc. After that, the nutritional information for a portion of the identified food is retrieved from a database and displayed to users. The app main features include:

- Taking images of food
- Recognizing food elements from the figure
- Retrieve nutritional information from database
- Information displayed

The focus of the project is application of AI technologies, relevant API (Google AI APIs, etc.), and web apps technologies (Bootstrap, Angular, NodeJS, etc.)

Reference: https://www.caloriemama.ai/

PROJECT 2 – TOURISM STORY MAP WEBSITE

Story Map is a representation of your story that highlights the locations of a series of events. A typical Story Map is a zoomable map of a region. In the map, one can mark several locations where story event happens. When clicking on each of the marks, a detailed view of the event should be displayed. The detail of the event can be a photo, an article, a clip, or any multimedia that attaches to the event.

The project's objective is to create a story map for tourists traveling to Seville, Spain. Seville is the capital of southern Spain's Andalusia region. It's famous for orange trees, flamenco, and bullfighting. We would like to have a map with 15 top attractions in Seville (according to tripadvisor.com). The Story Map should also suggest an initiative for a 3-day trip and 5-day trip in Seville.

Must-have features:

- A main view of a zoomable map of the location
- Zoom in and zoom out the map with sufficient details
- Marks of 15 top attractions of the location (according to tripadvisor)
- A detailed view of each attraction
- Tourist information for each attraction: description, how to get there, entrance fee, number of visitors, any notices, etc.
- A detailed view shows at least three best photos of the attraction
- A detailed view contains also a description of the attraction
- An easy way to change content for each attraction detail
- A page to suggest trips in the location
- Beautiful and interactive user experience
- Mobile-first design
- Two different ways of showing the map in mobile and laptop views

Nice-to-have features:

- Traffic tracking – count and show how many times an attraction is clicked in the map
- Add new marks in the map
- Add/edit/remove content for each attraction
- Display and add video clips to each attraction

- Login/logout
- Trip management: add/remove attraction into an initiative trip
- Embedding to Facebook

Possible technologies:

- Javascript, HTML5, CSS3
- React JS, Storymap JS
- PhP

Output reference:

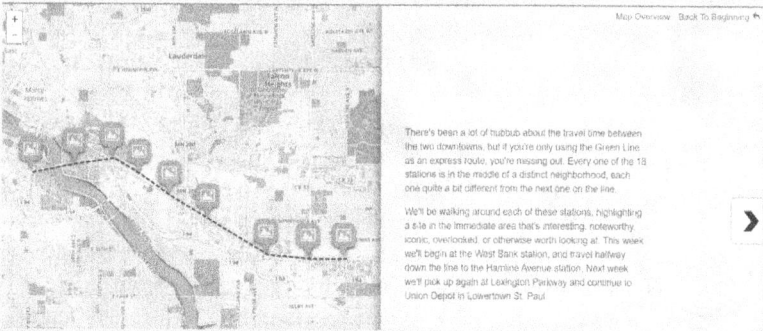

Link: https://www.minnpost.com/stroll/2014/06/hockey-hip-hop-and-other-green-line-highlights/ (Lenker til en ekstern side.)

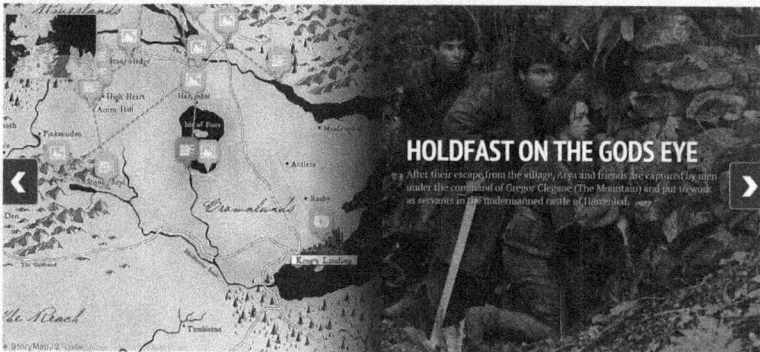

Link: https://storymap.knightlab.com/examples/aryas-journey

PROJECT 3: PERSONAL TRAVEL PLANNING APPLICATION

Open data fuels the development of new and innovative ICT solutions for the transport sector. The goal of the project is to develop a personal travel planning application using open transport data. The application will provide support for door-to-door travel in major Norwegian cities, e.g., Oslo and Trondheim. Various transport modes and services will be considered, e.g., bus, train, personal car, taxi, car sharing, and city bike. The application will support user with functionalities such as to specify the travel information and preferences; select from alternative travel itineraries; give notifications and feedback during travel, e.g., delays, traffic queue, when and where the user shall go/drive; suggest changed itineraries when deviation occurs; show parking places and availability if personal car is used; locate and reserve bikes when city bike service is used.

The project assignment is associated with the NFP project "Open Transport Data," which started in April 2016 with a duration of 3 years, and Statens vegvesen, SINTEF, ITS Norge, Kystverket, and Oslo kommune are partners in the project. The "Open Transport Data" project has established a prototype federated catalogue for open transport data in Norway. The personal travel planning application will find the needed open data using the services provided by the federated catalogue (such as the semantic search) and automatically switch to local open data when the location changes (e.g., when using the city bike services and parking services).

The students can choose programming languages. Web and mobile development skills are preferred, e.g., HTML 5, JavaScript, JSON, Java, Android. The code will be open source and the documentation will be in English.

CONTACT DETAILS:

Tasks (expectations)

- Survey existing approaches and applications identify what is good and what is missing.
- Define the desired features and requirements and implement them, using open transport data as starting point.

- Evaluate how open data can contribute to innovations from the experience of this process: what data is missing and should be open, what can be improved for existing open data (e.g., data format, API, description), what can improve the search and use of open data (e.g., will semantic search help find desired data more easily and accurately).

Resources and links

- *Existing examples*
- *Open transport data*

PROJECT 4: AADHAAR BIOMETRIC-BASED AUTOMATIC DOCUMENTLESS TRAFFIC MANAGEMENT SYSTEM

This project was a sociotechnological project, the idea of which was derived from the real-life problem of carrying hard copies of the original documents related to vehicles. The idea was to allow the verification of the driving-related documents such as driving license, vehicle registration, vehicle pollution certificate, vehicle insurance cover, etc., by the traffic police officer without the need for the vehicle driver to carry such documents physically. The problem domain was much explored as a result of personal experiences and discussions with the officials. This made it easy for the project team to straightway implement the project using software engineering approaches with higher level of confidence and certainties. Although the project was innovated later on by other students (please refer to Chapter 6), yet this project was the very niche and innovate product that had its real applicability in the real life.

This project has following advantages from software engineering point of view:

- Well-explored problem domain
- Stable requirements
- Low uncertainty with market
- High accuracy in planning
- High programming experience of team members
- Support of the software industries (as the student team member was on internship that time)

This software provides following functional utilities:

- On-the-spot challans and data update accordingly
- Online verification of vehicle-related documents using Aadhar number and/or figure print
- Possibility of vehicle owner to allow driving of the vehicle by authorized person (for instance, his son), thereby allowing identification of vehicle theft, etc.

The project code was filed for copyright to the Registrar of Copyrights, India, which was granted vide number 10065/2016-CO/SW.

The car driver needs to carry various documents such as Driving License, Car RC Copy, and pollution certificate while driving. The documents can be checked by traffic police any time during the course of journey and may be used and seized as security during traffic challans. This presents few limitations and opportunity for the traffic department as under:

- The complete process of release of seized document is time-consuming and effortful because the challan needs to be deposited to get back the document.
- In case the driver forgets to bring the documents (which he possesses) along with him during journey, then he will be issued a challan needlessly.

The opportunities are as follows:

- Allowing hassle-free documentless driving by providing at the spot online document verification system.
- Allowing hassle-free challan and payment facility, thereby making the system of document seizure and release unnecessary.
- Automatic storage of challan data, which then can be subjected for necessary analysis by higher management. For example, if maximum challans of yellow light jumping are at Sector 12 of the city, the management can see if it is due to error in timing LED display.
- The online embedded system will be developed to provide at the spot online system that allows the police to verify the

documents and even issue documentless challan. The challan information can be analyzed by higher management for necessary corrective measures.

The working of the Internet-enabled Aadhaar biometric-based system is as follows:

- The driver is supposed to enter his Aadhaar number and biometric, which then is authenticated online.
- After successful authentication, there are four options:
- View documents: The traffic police can see the complete details of car documents including the details of owner, photo, and details. The details of existing pollution certificate can also be seen.
- Issue challan: Once details of challan are entered in system, an entry is made against the details of the car driver. If an attempt is made to view documents later on by another policemen, then it will show that driver was challenged (which still is not paid), along with details of challan, etc.
- View history: Under this option, policeman can make an access to all previous driving licenses, pollution certificates, and RCs. Driving license and RC are renewed after long time period and may be issued in duplicate. Pollution certificate is to be obtained frequently after 3 or 6 months.
- Challan payment: Payment can be collected at the spot and corresponding information will be entered in the system. In case the driver takes the option of payment later either at police collection center or online, the same information will be entered in the system and can be viewed any time.
- The higher management can analyze the traffic challan for complete year, month, week, or a day for whole city, sector, or small region. The pending payment for issued challan can also be seen.

Support required from other departments:

- The RC and Driving License issuing department entered details will be shared online with the proposed system.

- The challan issue and payment information are to be linked to the system.
- Pollution issuing certified centers must be linked online so that the pollution issued certificates information can be linked to the system.
- Payment gateways must be linked to the system.

The project was innovated in next semester in following ways:

- New functionality related to tracing of the location of the vehicle stole by the thefts
- New functionality to detect accidents and informing nearby hospitals about the same
- New research algorithm to detect accidents

Appendix 2
Student background questionnaire

STUDENT BACKGROUND QUESTIONNAIRE

Prior experience with programming

- Front-end development: _____
- Back-end development: _____
- Database: _____
- Advanced technologies: _____

Prior experience with teamwork:

- Large team-size: _____
- Teamwork approach: _____
- Experience with leading: _____
- Experience with coordinating: _____

Expectation with the course:

- Gained experience: _____
- Ambition: _____

Appendix 3
Lectures in PRO1000, Spring semester 2020

LECTURES	DESCRIPTIONS
Lecture 1 – Introduction to project management	Brief overview of project management areas
	Customer presenting their projects
Lecture 2 – Project planning	Scope, time, risk, communication planning
	Students presenting their understanding about problem domains
Lecture 3 – Project execution 1	Agile process and practices 1
Lecture 4 – Project execution 2	Teamwork
	Students presenting their understanding about solution domains
Lecture 5 – Project execution 3	Agile process and practices 2
Lecture 6 – Practical topics	For example, details about GitHub, Testing, Security, Big Data, etc., with hands-on examples. Typically, a guest lecture
	Students present their experience
Lecture 7 – Project closing	Project close-up
	Presentation skills
	Demo day

Appendix 4
Suggested structure of the project report

Cover page
Abstract
Indexing pages
Chapter 1 - Introduction
Chapter 2 - Project planning
Chapter 3 - Prestudy of the solution space
Chapter 4 - Software development method
Chapter 5 - Product backlog
Chapter 6 - Architectural design
Chapter 7 - Testing
Chapter 8 - Sprint summaries
Chapter 9 - Project evaluation
References

Appendix 5
Templates of Sprint reports

Meta information

[About your teams/your projects]

Sprint goals

[The goal section states **why** it is worthwhile to undertake the sprint. Typically one goal should be given]

[Bad examples: To complete all four stories that we scheduled for implementation. Good examples: In this Sprint, we allow users to see the list of lawyers in the homepage]

Definition of done

[a shared understanding of what it means for work to be complete Example:

- DoD of each single user story included in the Sprint is met
- To-do list is completed
- All unit tests passing
- Product backlog updated
- Project deployed on Mac OS
- Tests on laptops, iPhone, and Android phones
- The performance tests passed
- All bugs fixed
- Sprint marked as ready for the production deployment by the product owner

]

Sprint backlog items

[The sprint backlog is a list of tasks identified by the Scrum team to be completed during the Sprint]

Kanban/Trello board
[You can print the screenshot of the kanban board here]

Summary of Sprint review meeting with customer signature
[the Scrum team shows what they accomplished during the sprint. Typically this takes the form of a demo of the new features.]
Example:
Participants: Member1, Member2, Member3
Meeting duration, location:
Summary of what is discussed
Demonstration result

ITEM ID	ITEM DESCRIPTION	ESTIMATION	STATUS	DEMO
U01	As a site member (?), I can scroll through a listing of jobs. (There won't be enough at first to justify search fields.)	8	Finished	Yes
U02	As someone who wants to hire, I can post a "help wanted ad".	20	We started but so many open issues	No
U03	As a site admin, I can edit and delete help wanted ads.	15	Finished	No
U04	As a site visitor, I want to be able to read some of your articles.	5	Untouched	No

Summary of customer feedback:
- Story U01 is complete. The customer is pleased with the user interface
- Story 02 was misunderstood. The customer clarified the issue and led to a new user story
- Requirement on back end is negotiated and customer agreed with new scope.

Summary of Sprint retrospective meeting
Participants: Member1, Member2, Member3
Meeting duration, location:
Summary of what:
- Start doing
- Stop doing
- Continue doing

References

Albanese, M. A., & Mitchell, S. (1993). Problem-based learning: A review of literature on its outcomes and implementation issues. *Academic Medicine*, 68(1), 52–81. https://doi.org/10.1097/00001888-199301000-00012

Bacon, D. R., Stewart, K. A., & Silver, W. S. (2016). Lessons from the best and worst student team experiences: How a teacher can make the difference. *Journal of Management Education*. https://doi.org/10.1177/105256299902300503

Bell, S. (2010). Project-based learning for the 21st century: Skills for the future. *The Clearing House. A Journal of Educational Strategies, Issues and Ideas*, 83(2), 39–43. https://doi.org/10.1080/00098650903505415

Biggs, J., & Tang, C. (2011). *Teaching for Quality Learning at University*. McGraw-Hill Education, UK.

Blumenfeld, P. C., Soloway, E., Marx, R. W., Krajcik, J. S., Guzdial, M., & Palincsar, A. (1991). Motivating project-based learning: Sustaining the doing, supporting the learning. *Educational Psychologist*, 26(3–4), 369–398. https://doi.org/10.1207/s15326985ep2603&4_8

Boehm, B. W. (1991). Software risk management: Principles and practices. *IEEE Software*, 8(1), 32–41. https://doi.org/10.1109/52.62930

Bourque, P., Fairley, R. E., et al. (2014). *Guide to the Software Engineering Body of Knowledge (SWEBOK (r)): Version 3.0*. IEEE Computer Society Press.

Brooks, F. P. (1995). *The Mythical Man-Month* (Anniversary ed.). Addison-Wesley Longman Publishing Co., Inc.

Bruegge, B., Krusche, S., & Alperowitz, L. (2015). Software engineering project courses with industrial clients. *Transactions on Computing Education*, 15(4), 17:1–17:31. https://doi.org/10.1145/2732155

Chen, L. (2015). Continuous delivery: Huge benefits, but challenges too. *IEEE Software*, 32(2), 50–54. https://doi.org/10.1109/MS.2015.27

Feldt, R., Höst, M., & Lüders, F. (2009). Generic Skills in Software Engineering Master Thesis Projects: Towards Rubric-Based Evaluation. 22nd Conference on Software Engineering Education and Training, 12–15. https://doi.org/10.1109/CSEET.2009.54

Fitzgerald, B., & Stol, K.-J. (2014). Continuous Software Engineering and Beyond: Trends and Challenges Categories and Subject Descriptors. RCoSE 2014 Proceedings of the 1st International Workshop on Rapid Continuous Software Engineering, 1–9. https://doi.org/10.1145/2593812.2593813

Fitzgerald, B., & Stol, K. J. (2017). Continuous software engineering: A roadmap and agenda. *Journal of Systems and Software*, 123, 176–189. https://doi.org/10.1016/j.jss.2015.06.063

Goodman, P. S., & Leyden, D. P. (1991). Familiarity and group productivity. *Journal of Applied Psychology*, 76(4), 578–586. https://doi.org/10.1037/0021-9010.76.4.578

Gupta, V., Chauhan, D. S., & Dutta, K. (2015). Exploring reprioritization through systematic literature surveys and case studies. *SpringerPlus*, 4(1), 539.

Gupta, V., Fernandez-Crehuet, J. M., & Hanne, T. (2020a). Freelancers in the software development process: A systematic mapping study. *Processes*, 8(10), 1215.

Gupta, V., Fernandez-Crehuet, J. M., Hanne, T., & Telesko, R. (2020b). Fostering product innovations in software startups through freelancer supported requirement engineering. *Results in Engineering*, 8, 100175. https://doi.org/10.1016/j.rineng.2020.100175

Gupta, V., Fernandez-Crehuet, J. M., Hanne, T., & Telesko, R. (2020c). Requirements engineering in software startups: A systematic mapping study. *Applied Sciences*, 10(17), 6125.

Helle, L., Tynjälä, P., & Olkinuora, E. (2006). Project-based learning in post-secondary education – Theory, practice and rubber sling shots. *Higher Education*, 51(2), 287–314. https://doi.org/10.1007/s10734-004-6386-5

Heywood, J. (2000). *Assessment in Higher Education: Student Learning, Teaching, Programmes and Institutions*. Jessica Kingsley.

Hoegl, M. (2005). Smaller teams–better teamwork: How to keep project teams small. *Business Horizons*, 48(3), 209–214. https://doi.org/10.1016/j.bushor.2004.10.013

Hyman, B. (2001). From capstone to cornerstone: A new paradigm for design education January. *International Journal of Engineering Education*, 17(4), 416–420.

ISO/IEC/IEEE std 24765. (2010). Scrum Guide. https://www.scrumguides.org/scrum-guide.html#purpose

Jaakkola, H., Henno, J., & Rudas, I. J. (2006). IT Curriculum as a Complex Emerging Process. IEEE International Conference on Computational Cybernetics, 1–5. https://doi.org/10.1109/ICCCYB.2006.305731

Jacobson, I., Lawson, H. "Bud", Ng, P.-W., McMahon, P. E., & Goedicke, M. (2019). *The Essentials of Modern Software Engineering: Free the Practices from the Method Prisons!* Association for Computing Machinery and Morgan & Claypool.

Kruchten, P. B. (1995). The 4+1 View Model of architecture. *IEEE Software*, 12(6), 42–50. https://doi.org/10.1109/52.469759

Lethbridge, T. C. (1998). The relevance of software education: A survey and some recommendations. *Annals of Software Engineering*, 6(1), 91–110. https://doi.org/10.1023/A:1018917700997

Mahnič, V., & Hovelja, T. (2012). On using planning poker for estimating user stories. *Journal of Systems and Software*, 85(9), 2086–2095. https://doi.org/10.1016/j.jss.2012.04.005

Nguyen-Duc, A., & Abrahamsson, P. (2016). Minimum Viable Product or Multiple Facet Product? The Role of MVP in Software Startups. In H. Sharp, & T. Hall (Eds.), *Agile Processes, in Software Engineering, and Extreme Programming* (pp. 118–130). Springer International Publishing.

Nguyen-Duc, A., Jaccheri, L., & Abrahamsson, P. (2019). An Empirical Study on Female Participation in Software Project Courses. 2019 IEEE/ACM 41st International Conference on Software Engineering: Companion Proceedings (ICSE-Companion), 240–241. https://doi.org/10.1109/ICSE-Companion.2019.00094

Nguyen-Duc, A., Münch, J., Prikladnicki, R., Wang, X., & Abrahamsson, P. (Eds.) (2020). *Fundamentals of Software Startups: Essential Engineering and Business Aspects*. Springer International Publishing. https://doi.org/10.1007/978-3-030-35983-6

Nguyen-Duc, A., Seppänen, P., & Abrahamsson, P. (2015). Hunter-gatherer Cycle: A Conceptual Model of the Evolution of Software Startups. Proceedings of the 2015 International Conference on Software and System Process, 199–203. https://doi.org/10.1145/2785592.2795368

Petkov, D., & Petkova, O. (2006). Development of scoring rubrics for IS projects as an assessment tool. Issues in Informing Science & Information Technology, 3, 499–511.

PMBOK. (2013). A Guide to the Project Management Body of Knowledge: PMBOK(R) Guide (5th ed.). Project Management Institute.

Ries, E. (2011). The lean startup: How today's entrepreneurs use continuous innovation to create radically successful businesses, Crown Business, 2011. 320 pages.

Saqr, M., Nouri, J., & Jormanainen, I. (2019). A Learning Analytics Study of the Effect of Group Size on Social Dynamics and Performance in Online Collaborative Learning. In M. Scheffel, J. Broisin, V. Pammer-Schindler, A. Ioannou, & J. Schneider (Eds.), Transforming Learning with Meaningful Technologies (pp. 466–479). Springer International Publishing. https://doi.org/10.1007/978-3-030-29736-7_35

Schraw, G., Dunkle, M. E., & Bendixen, L. D. (1995). Cognitive processes in well-defined and ill-defined problem solving. Applied Cognitive Psychology, 9(6), 523–538. https://doi.org/10.1002/acp.2350090605

Schwalbe, K. (2013). Information Technology Project Management, Revised (7th ed.). CENGAGE Learning: UK.

Schwalbe, K. (2019). Information Technology Project Management (9th ed.). CENGAGE Learning: Boston.

Sedelmaier, Y., & Landes, D. (2014). Software Engineering Body of Skills (SWEBOS). IEEE Global Engineering Education Conference (EDUCON), 395–401. https://doi.org/10.1109/EDUCON.2014.6826125

Sutherland, J. V., Patel, D., Casanave, C., Miller, J., & Hollowell, G. (Eds.) (1997). Business Object Design and Implementation: OOPSLA '95 Workshop Proceedings 16 October 1995, Austin, Texas. Springer-Verlag. https://doi.org/10.1007/978-1-4471-0947-1

Taylor, R. N., Medvidovic, N., & Dashofy, E. M. (2009). Software Architecture: Foundations, Theory, and Practice (1st ed.). Wiley.

Tuckman, B. W. (1965). Developmental sequence in small groups. Psychological Bulletin, 63(6), 384–399. https://doi.org/10.1037/h0022100

Tuckman, B. W., & Jensen, M. A. C. (1977). Stages of small-group development revisited. Group & Organization Studies. https://doi.org/10.1177/105960117700200404

Valstar, S., Krause-Levy, S., Macedo, A., Griswold, W. G., & Porter, L. (2020). Faculty Views on the Goals of an Undergraduate CS Education and the Academia-Industry Gap. Proceedings of the 51st ACM Technical Symposium on Computer Science Education, 577–583. https://doi.org/10.1145/3328778.3366834

Wieringa, R., Maiden, N., Mead, N., & Rolland, C. (2006). Requirements engineering paper classification and evaluation criteria: a proposal and a discussion. Requirements engineering, 11(1), 102–107.

Index

Note: **Bold** page numbers refer to tables and *italic* page numbers refer to figures.

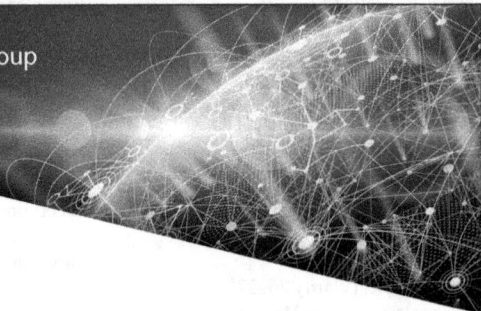

For Product Safety Concerns and Information please contact our EU
representative GPSR@taylorandfrancis.com
Taylor & Francis Verlag GmbH, Kaufingerstraße 24, 80331 München, Germany

www.ingramcontent.com/pod-product-compliance
Lightning Source LLC
Chambersburg PA
CBHW061331220326
41599CB00026B/5133

9 781032 002538